Global Warming
UNCHECKED

Global Warming
UNCHECKED
SIGNS TO WATCH FOR

Harold W. Bernard, Jr.

INDIANA UNIVERSITY PRESS
Bloomington & Indianapolis

The paper used in this publication meets the minimum requirements of American
National Standard for Information Sciences—Permanence of Paper for Printed
Library Materials, ANSI Z39.48-1984.
∞™

Manufactured in the United States of America

Library of Congress Cataloging-in-Publication Data

Bernard, Harold W., Jr.
    Global warming unchecked : signs to watch for / Harold W. Bernard, Jr.
        p.   cm.
    Includes index.
    ISBN 0-253-31178-0. — ISBN 0-253-20819-X (pbk.)
    1. Global warming.   2. Global warming—Social aspects—United
States.   3. United States—Climate—Environmental aspects.
    I. Title.
QC981.8.G56B45   1993
363.73'87—dc20   92-41065

1 2 3 4 5 97 96 95 94 93

For my grandson, Nicholas Andrew Dracos.
Sorry about the environment we've left you, Nicholas.
"Spell my name right, Gramps," he said.

Contents

Preface

"You doomsayers are always wrong," one critic of the greenhouse effect railed at me as we discussed global warming. The critic was unwilling to concede that maybe "doomsayers" are "always wrong" because sometimes forestalling actions are taken based on their warnings.

For that matter, I'm not willing to kowtow to the charge that predictions of "doom" are always wrong. Consider the threat of unchecked population growth. China did, and it took drastic steps to curtail its population explosion. Bangladesh didn't, and as a result millions of people in that overcrowded nation are forced to subsist on land barely above sea level. Sometimes they don't subsist very long. Periodic floods and cyclones snatch tens of thousands of lives at a shot. That's death on an immense scale. So immense, we in the United States can't even imagine it. But if you're a Bangladeshi, the doom is real.

But doom is not the theme of this book. The theme is challenges. Norwegian Prime Minister Gro Harlem Brundtland, who chairs the United Nations World Commission on Environment and Development, speaking of the greenhouse effect, points out, "The impact of climate change may be greater and more drastic than any challenges mankind has faced with the exception of nuclear war." U.S. Vice President Albert Gore sees the challenge of the greenhouse effect, too. "It's the most serious environmental problem we've ever faced," he warns.

The challenges go beyond the environment, however. In the United States they are intertwined with our energy future and our economic viability. In short, by addressing the greenhouse threat we declare our energy independence, maintain our position as a world economic power, and become a trailblazer on a path leading to a cleaner, more energy-efficient world. But perhaps that's too utopian, too hard to envision.

But so too is a drastically warmer world. It's difficult for me, as a meteorologist, to believe that we really could warm our planet as much and as rapidly as is forecast by greenhouse-effect researchers. Yet the results are there, and on balance I find them convincing.

That's what frightens me. We're performing an experiment—albeit an inadvertent experiment—of unprecedented proportions with our atmosphere. The results, to borrow a teenage phrase, could be awesome, and in some cases devastating. If I err somewhat in sounding too much like a doomsayer in this book, the breach of balance is intentional. I am concerned. We do need to take action. The stakes are extraordinarily high, and there will be no second chance.

It's easy for critics to sit back, view the greenhouse threat academically, and fume that there is no proof and thus no problem. But merely saying so does not make it so. The consequences of a greenhouse world are not academic. They are real. And they are enormous.

If we care about our future, we've no choice but to act now.

Several people have earned my special thanks for their assistance in the preparation of this book: Harold Bernard (senior), my father and an author in his own right; David Spiegler, president of DBS Associates and Impact Weather; and Stuart Soroka, a former TV meteorologist. These three individuals reviewed the entire manuscript and helped make it a better book.

My father provided the layman's viewpoint, Spiegler played the devil's advocate, and Soroka critiqued my meteorology and writing. My father is an educational psychologist, Spiegler a businessman, researcher, and Certified Consulting Meteorologist. Soroka was a good, practical weatherman. Tragically, Stuart died before the book came to print, but he believed in it as much . . . no, almost more than I did. Thanks, Stuart. I miss you.

And thanks, too, to my wife, Christina, who found most of the grammar mistakes I made.

A large number of people and organizations contributed time and information to this effort, either by answering my questions over the telephone or sending me information, sometimes both. The organizational affiliations of the contributors in the following list are as of the time I corresponded with them: Ernie Abrams, GTE; John Adams, University of Maryland; Leon Hartwell Allen, Jr., University of Florida; Larry Dozier, Central Arizona Project; Kerry Emanuel, Massachusetts Institute of Technology; Lawrence Gates, Oregon State University; Peter Gleick, Pacific Institute; James Hansen, NASA Goddard Space Flight Center; Richard Heim, National Climatic Data Center; Steve Jenkins, Arizona Department of Natural Resources; John Jansen, Southern Company Services; Wally Jones, EPA (Atlanta office); Pete Leavitt, Weather Services Corporation; Syukuro Manabe, Geophysical Fluid Dynamics Laboratory; Patrick McIntosh, Environmental Research Laboratories; Scott Miller,

EPA (Atlanta office); National Safety Council; Stephen Schneider, National Center for Atmospheric Research; John Topping, Climate Institute; and Paul Waggoner (retired), Connecticut Agricultural Experiment Station.

PART I

The Immediate Future

1 / The Warning of the 1980s

The greenhouse effect . . . global warming. The wolf isn't at our door, he's in. And his breath is hot and dry. We've had plenty of warning, but how many people have been listening? Only a handful, apparently.

Crying Wolf

By the end of 1989—a year in which the remarkable U.S. drought and heat of the previous year had disappeared—most people were, in fact, listening to the greenhouse-effect contrarians. As quickly as the media had jumped aboard the greenhouse express, they hopped off to tell everyone (often not very factually) they really hadn't missed the train . . . that the wolf really wasn't in the door; that the greenhouse-effect believers had only been crying wolf.

What people were hearing was that since the country, in 1989, didn't dry up and blow away, or sizzle under unrelenting sunshine, warnings of the greenhouse effect had been overblown. Not true. The fact is we will not move toward a greenhouse world in a steady, straight-line fashion with each year growing hotter and, in some regions, drier. There will still be years in the near future featuring no sustained drought or heat.

In fact, an event that occurred in mid-1991 is likely to temporarily reverse the overall global warming trend of recent years. On June 15, 1991, the eruption of Mount Pinatubo in the Philippines blew massive amounts of ash and gas, including sulfur dioxide, into the atmosphere. Sulfur dioxide is important because it is transformed into tiny droplets of sulfuric acid and becomes a volcanic aerosol. Volcanic aerosols, upon reaching the upper part of the atmosphere, act as sunlight reflectors. The net result is that less sunlight reaches the surface of the earth and global temperatures fall slightly. In the case of Mount Pinatubo—whose eruption spewed out the greatest volume of aerosols this century—scientists expect general cooling of the earth through 1993, with greenhouse warming resuming thereafter.

The real danger is that such cooling will only mask what I feel is the most immediate climatic threat to the United States: an absolutely devastating drought and heat wave in the midsection of the nation before the end of this century.

By the end of 1989, however, such concerns were being widely dismissed. For instance, it was pointed out that there had been no significant increase in U.S. temperatures this century. True. But the trouble with embracing that point as evidence of the lack of a greenhouse effect is that the United States represents only a tiny fraction of the earth's surface. The greenhouse effect is a global phenomenon. On a global scale, temperatures *have* risen over the past hundred years; for instance, on a global scale 1990 was the warmest year in recorded history!

Still, in early 1990 a newspaper headline proclaimed, "Earth-warming trend fails to show up on satellite data, scientists say." The gist of the accompanying story was that no net global warming (or cooling) had been detected by weather satellites from 1979 through 1988. But a ten-year period does not necessarily offer any insight into longer trends. As one scientist involved in the satellite-based research, John Christy of the University of Alabama-Huntsville, cautioned, "There is no guarantee that if you take a ten-year segment out of a long time that you'll get the overall trend." In fact, most of the recent worldwide warming occurred between about 1965 and 1980. During the 1980s the upward trend slowed, albeit at a very high plateau: the 1980–89 decade was the warmest in over a century!

As the 1990s began, some writers of popular science were telling us that increasing cloudiness, fostered by increasing evaporation due to increasing greenhouse-effect warmth, would stem or reverse the warming. Not so. You can't have cooling *and* warming from the same cause. Worldwide, you get one or the other. From the greenhouse effect you get warming. There is no argument on that point. More clouds may slow the warming, but they will not stop or reverse it.

Other writers were touting the fact that certain recent computerized climate model runs—using more sophisticated representations of clouds—were predicting less greenhouse warming than before. Clouds are only part of the greenhouse-effect puzzle, however, and different climate models (and there are several) put the puzzle together differently. Since 1980 the generally accepted range of answers given by these models has expanded slightly, *but more at the upper end than at the lower end of the warming scale.*

And of course there were those who told us the models weren't any good at all . . . that they hadn't predicted the degree of greenhouse warming that should have occurred over the past 130 years. In fact, the amount of warming due to the greenhouse effect would have been too small to detect until now. And even now the amount of global warming directly attributable to the greenhouse effect is debatable.

But we do know a couple of things for certain. Some greenhouse-effect climate models have replicated quite accurately global temperature trends since 1958. (That year was chosen as a starting point, since that was when we first had precise measurements of gases contributing to the greenhouse effect.) The same models warn that starting about now, the greenhouse effect will begin to overwhelm everything else, every other climate control.

Fred Wood of the Office of Technology Assessment of the United States Congress, writing in the conservative *Bulletin of the American Meteorological Society,* says it another way: "With respect to longer-term climate trends, during the span of several centuries or more . . . data . . . suggest that the recent warming is still within—although perhaps pushing—the upper limits of natural variability." Wood offers a veiled warning: "In sum, the state of the global climate appears to be at a critical juncture, with some current indicators at or close to historic limits."

But we've had other warnings.

The First Warning

As far back as the late 1890s there was concern over the release of carbon dioxide (CO_2) into the atmosphere from the combustion of fossil fuels. Late last century the concern was over coal, which was being burned in massive quantities to fuel the Industrial Revolution. The individual most concerned was a Swedish chemist, Svante Arrhenius. He made the startling prediction that a doubling of atmospheric CO_2 would lead to worldwide warming on the order of 9° F (5° C). His was the very first warning of what is now popularly known as the "greenhouse effect."

But Arrhenius was far ahead of his time, and his dire forecast was met with scorn by the scientific community. It was not until after the middle of this century that significant interest in the effects of accumulating atmospheric CO_2 regenerated.

Contemporary Warnings

In the mid-1960s, researchers brushed the dust off Arrhenius's work and decided it was time for an updated look. In 1967 Syukuro Manabe and

Richard Wetherald, of the U.S. Department of Commerce's Geophysical Fluid Dynamics Laboratory at Princeton University, issued a new warning: expect a global temperature increase of slightly more than 4° F (2° C) if atmospheric CO_2 is allowed to double from its preindustrial level.

Since then there has been an avalanche of research on the greenhouse effect. The consensus among scientists now is that a doubling of atmospheric CO_2 will indeed lead to a warmer Earth, somewhere in the range of 3° to 8° F (1.5° to 4.5° C).

By the late 1970s, greenhouse-effect alarms were sounding publicly. In its May 3, 1979, issue, the respected British scientific journal *Nature* put the subject into harsh perspective: "The release of carbon dioxide to the atmosphere by the burning of fossil fuels is, conceivably, the most important environmental issue in the world today."

In October 1983 both the U.S. Environmental Protection Agency and the National Research Council issued reports on the greenhouse effect. The reports made newspaper and television headlines, but not for long. The newsmaking potential of the reports was quickly obliterated by the deadly terrorist attack on the U.S. Marines barracks in Beirut, Lebanon (October 23), and the U.S. "miniwar" on the Caribbean island of Grenada (October 25).

Probably in the long run it wouldn't have made any difference had the news of the greenhouse effect lingered in the public eye. The overall attitude of people seems to be "why should we worry about CO_2 when there's AIDS, poverty, and war?" Thomas Schelling, a professor of political economy and a public policy specialist at Harvard's John F. Kennedy School of Government, puts it more bluntly. "Even if the worst of the predicted climate changes show up," he says, "carbon dioxide isn't going to be on my list of the half-dozen things we need to worry about."

U.S. industry, too, has been less than enthusiastic about addressing the atmospheric CO_2 problem. Industrial plants, particularly electric utilities, rely on CO_2-producing fossil fuels—coal, oil, and natural gas—for energy. Says Fritz Kalhammer of the Electric Power Research Institute, "I don't think anyone wants to cripple the economy of this country over a perceived problem that is not even well understood."

The bottom line is that while greenhouse-effect reports and articles have made good reading and spectacular headlines, the general attitude of the American public—and worse, of American politicians (with the exception of a precious few)—has been: "Ho hum. Another global crisis. Well, it'll just have to wait its turn."

Guess what. It didn't wait. It's here. The wolf is in the door.

Early in 1988 James Hansen, a climatologist and director of the

Global Means

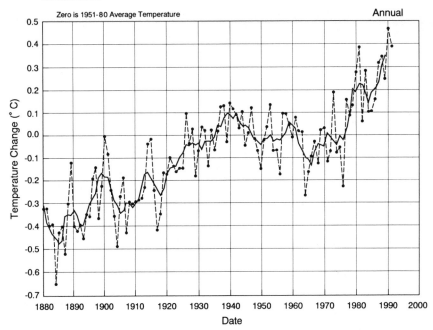

FIGURE 1.1 Global temperature trend since 1880. Reproduced, with permission, from James Hansen and Helene Wilson, NASA Goddard Institute for Space Studies, "GISS Analysis of 1991 Global Surface Air Temperature," Jan. 6, 1992.

National Aeronautics and Space Administration's Goddard Institute for Space Studies, told a U.S. Senate panel that current evidence presents a very strong case "that the greenhouse effect has been detected, and it is changing our climate now." He said that with "99 percent confidence" we can state that recent global warming (see figure 1.1) is a "real warming trend," not one occurring by chance or accident.

Something else is apparently not occurring by chance or accident: a glimpse of future climate. Certain weather events of the 1980s may have given us truncated yet palpable warnings of things to come . . . things to come in the near future. The 1990s—after the effects of Mount Pinatubo have disappeared—may not be as kind to us as the 1980s. The wolf's breath is hot and dry.

The Warning of 1980

July and August 1980 brought the first of a series of great droughts and sizzling heat waves that would stalk the United States during the remainder of the decade. The core of the 1980 heat scorched the land from northeast Texas northward into Kansas and Missouri and eastward through Arkansas. Rainfall during July and August was less than 50 percent of normal for most of eastern Texas, Oklahoma, Arkansas, and Louisiana.

At the Dallas–Fort Worth airport an unprecedented string of days with maximum temperatures of 100° F (38° C) or more began on June 23. It was not until August 4, after forty-two consecutive days, that the temperature failed to reach or top 100° F. Along the way, Dallas–Fort Worth twice established a new high temperature record: 113° F (45° C) on June 26 and 27. In parts of Texas, Oklahoma, Kansas, and Missouri, the summer of 1980 was the hottest ever known. Hotter even than the summers of the great Dust Bowl days of the 1930s, although the heat then was centered farther north.

Occasionally, fiery fingers of 1980's intense hot weather crept into other areas of the country. New all-time records for heat were set at a number of locations in the Deep South on July 10 and 11: Macon, Georgia, 108° F (42° C); Meridian, Mississippi, 107° F (42° C); Pensacola, Florida, 106° F (41° C); and Atlanta, Georgia, 105° F (41° C). In Washington, D.C., the average monthly temperature for July was the highest in 110 years of record keeping: 82.3° F (27.9° C).

The heat and drought of 1980 were killers. They contributed to 1,200 deaths and to an economic disaster in the form of crop and livestock losses estimated at twenty billion dollars, mostly on the southern Plains.

There was another weather event in 1980 that was perhaps a harbinger of the future, but it was lost in headlines that touted scorched and parched earth. Early in August, Hurricane Allen swept through the Caribbean, tore across the western Gulf of Mexico, and careened into Texas just north of Brownsville. Although Allen was only a category three storm (on a scale of one to five) at landfall, prior to that time it had been the second most intense hurricane ever observed in the Atlantic–Caribbean–Gulf of Mexico area. (Just eight years later, in September 1988, a hurricane even more violent than Allen, following a similar track, laid waste certain Caribbean islands and part of Mexico. Chapter 3 discusses the greenhouse effect, hurricanes, and implications for our future.)

The Warning of 1983

The years 1981 and 1982 were relatively kind to U.S. agriculture, but drought and heat soon returned. In the summer of 1983 much of the

United States again roasted under cloudless skies and unrelenting sunshine. July and August brought dry conditions similar in size to those of 1980 but centered somewhat farther north. The desiccation this time reached into the Canadian prairies, although the dusty heart of the drought—in August—was in a crescent-shaped area extending from west Texas northward to Nebraska and Iowa, and from there southeastward through Missouri and Tennessee to Georgia.

The blistering, rainless weather again evoked comparisons with the legendary Dust Bowl summers of the 1930s. Some observers thought that in terms of the *combination* of heat and drought, the summer of 1983 may have been worse than the sizzling summer of 1936. Many farmers in 1983 lost virtually all of their corn crop. Overall, summer crop production (especially corn and soybeans) was down by almost 30 percent, with losses totaling ten billion dollars. And as a direct result of the suffocating heat, 220 people died.

The Warning of 1984

Although 1984 and 1985 brought generally better weather for crops, a small but intense drought in 1984 gripped Montana and North Dakota and reached northward into southern Alberta and Saskatchewan. Wheat crop yields were down 20 percent in Canada, and losses exceeded one billion dollars.

Montana battled another facet of drought that year: fire. Governor Ted Schwinden said on August 18: "Our state literally is on fire from Glacier Park in the extreme northwest corner to the Custer [National] Forest and the Bull Mountains in the southeast." Nearly a quarter of a million acres (over 100,000 hectares) of forest and grasslands burned by the end of the month. Millions of dollars were spent fighting blazes that destroyed forty homes, forced the evacuation of hundreds of people, closed major highways, and damaged popular tourist areas.

The Warning of 1986

In 1986 the focus of weather headlines shifted to the Southeast. A January through July drought there was termed by the National Oceanic and Atmospheric Administration's Climate Analysis Center as "probably the worst in at least 99 years." From southern Virginia southward through the western Carolinas into parts of Georgia, Alabama, and Tennessee, crops withered and died, ponds and lakes dried up, and livestock and poultry died. Hay and corn crop losses alone exceeded one billion dollars.

In July the heat was unrelenting. From early July through early August Atlanta, Georgia, counted thirty-eight consecutive days with temperatures reaching or topping 90° F (32° C). The broiling heat peaked on July 21 when Macon, Georgia, hit 106° F (41° C); Columbia, South Carolina, and Augusta, Georgia, reached 105°F (41° C); and coastal Charleston, South Carolina, tagged 104° F (40° C).

The heat extended its tentacles to other parts of the South, too. On the last day of the month, Little Rock, Arkansas, tied its all-time maximum temperature record when the mercury rose to 110° F (43° C). The old record had been established during the great heat wave of 1936. (Chapter 2 compares that legendary heat wave—and others of the 1930s—with those of the 1980s and discusses the implications for our immediate future.)

The Warning of 1988

Unusually dry weather in the Pacific Northwest during 1987 forced the September mean flow of the mighty Columbia River to its lowest level in 108 years, but widespread drought didn't return to the United States until 1988. That year another great river, the Mississippi, suffered a similar fate.

The winter of 1986–87 was worthy of note on the northern Plains, but not for dryness. Across Montana, North and South Dakota, Minnesota, and Wisconsin, it was the mildest winter since 1895.

Earth-cracking drought came again in 1988. From the Pacific Northwest, across the northern Plains, through the midwestern cornbelt, and down into northern Georgia, the land once again was sucked dry of moisture. In the spring in rural Minnesota, clouds of dust rolled across the prairie, and even in broad daylight drivers were forced to turn on their automobile headlights. In Memphis, Tennessee, the Mississippi dwindled to its lowest level in over 100 years, and barge traffic came to a standstill. By mid-July 37 percent of the nation was suffering severe-to-extreme drought. (Across much of the Midwest and Plains only three months—June 1933, May 1934, and June 1936—had been drier than June 1988.)

The climax of the heat was reached on June 25. Temperatures soared to 100° F (38° C) or more from Michigan to Florida, and fifty cities set new high temperature records for the date. Cleveland, Ohio, established its all-time high mark, 104° F (40° C), as did Erie, Pennsylvania, with 100° F (38° C). Hot weather, to a lesser degree, continued through August.

By the time it was over, summer 1988—based on the magnitude and extent of heat—went into the record books as one of the four hottest of the century. (Only the Dust Bowl year of 1934 had a heat wave of greater proportions. The simmering summers of 1936 and 1952 were about equal to that of 1988.) Estimates pegged 1988 crop and livestock losses at fifteen billion dollars. Total direct economic losses and costs were near forty billion dollars.

The legacy of the 1988 drought didn't end with agriculture. Millions of Americans watched on television, in sadness and awe, as forest fires ran wild, consuming hundreds of thousands of acres in Yellowstone National Park during August and September. Dry weather and lightning contributed to the holocaust, but so did the park policy of not fighting naturally caused fires . . . at least until it was too late. (Is there a greenhouse-effect lesson for us here?)

Finally there was Gilbert . . . Hurricane Gilbert, termed by Robert Sheets, director of the National Hurricane Center, as "the granddaddy of all storms." Gilbert was the most intense storm ever observed in the Western Hemisphere. Its September track was close to that of 1980's Allen, but Gilbert thundered directly over Jamaica and the Yucatan Peninsula before weakening to a category three storm and crossing the Mexican coast just south of Brownsville, Texas. (In October, Hurricane Joan, a category four storm, crashed into Nicaragua, the first storm to do so since 1911.)

On a grander scale, a global scale, as James Hansen pointed out to the Senate panel in 1988, the four warmest years ever recorded had occurred in the 1980s (see figure 1.1). And 1987, he said, had been one of the earth's warmest years in modern times.

But as it turned out, 1988, 1990, and 1991 were even warmer, and 1989 wasn't far off the mark. Thus within a twelve-year period the eight warmest years in the past century had occurred. And 1990 was the hottest of them all. This conclusion was reached by researchers with the Meteorological Office of the United Kingdom and the University of East Anglia, as well as those of the National Aeronautics and Space Administration and the National Oceanic and Atmospheric Administration in the United States. It is clear the warnings didn't cease with 1988.

Parallels

Global temperatures in the 1980s had already exceeded those of the 1930s, generally conceded to be the warmest decade in recent times in the Northern Hemisphere. This is important, because parallels are often drawn between the climate of the 1930s and the greenhouse climate expected in the *near* future.

So, for the United States, consider the 1980s relative to the 1930s. The eighties brought four major onslaughts of heat and drought that triggered justified comparisons to the notorious Dust Bowl days of the thirties. (Again, the implications of this are discussed in chapter 2). The 1980s also brought two of the three most intense hurricanes ever seen in the Atlantic–Caribbean–Gulf of Mexico region. The third storm in this august group was a product of the 1930s: the Florida Keys Labor Day Hurricane of 1935 (more on this in chapter 3.)

Is It Really the Greenhouse Effect?

But are the climatic changes we're witnessing really a result of the greenhouse effect? For starters, consider the temperature records of James Hansen, the British, and others displaying the impressive global warming of recent years. Scientists say the probability of such changes occurring merely by chance are only one in 100.

W. Lawrence Gates, director of the Climatic Research Institute at Oregon State University, points out that on a global average basis these contemporary temperature rises "are not inconsistent with the general trend of [the greenhouse effect] models." In a cautious sort of way Gates is saying he agrees with Hansen.

But maybe it's not the greenhouse effect. Maybe the atmospheric scientists and climate modelers are wrong; after all, there are greenhouse-effect detractors and nonbelievers. Don't get your hopes up. Consider what Stephen Schneider, a climate modeler at the National Center for Atmospheric Research says: "The greenhouse effect is the least controversial theory in atmospheric science."

We are, in all likelihood, beginning to see CO_2-induced atmospheric warming emerge from what researchers call the "noise level" of global temperature fluctuations. That is, recent global temperature changes are becoming large enough that scientists can look at them and judge that something other than "natural" causes is driving the changes.

Actually, we'd probably better hope that recent global temperature trends *are* a manifestation of the greenhouse effect. If they aren't, if there's something else at work, if some other phenomenon we haven't detected is warming our climate, then we're in real trouble. Because if the greenhouse effect hasn't yet kicked in, when it does it will be starting from an even higher threshold, off on a running start to a warmer world.

2/ The Great Drought of the 1990s

I once worked as a weather forecaster with a private meteorological service in Connecticut. Included in our clientele were a number of radio stations. Sometimes when our forecast differed from that of the National Weather Service, a listener would call and chide us for not sticking with the "official" forecast. One of my co-workers, Ken Garee, had a good response to that. He would acknowledge that yes indeed, our forecast was different, and it certainly wasn't "official." "But," he added, "we'd rather be right than official." (We weren't always right, either.)

The picture presented in this chapter of near-term weather trends is not "official." It doesn't stem from the work of any research organization. It's simply my own speculation, based on such things as global temperature curves, climatic analogs, sunspot cycles, and some computer modeling results. But while it's not "official," I think it's right. I think it's something that's going to happen. Am I 100 percent certain that it will? Of course not. Am I 80 percent sure? No. But then I don't know how to quantify the probability. I'll plead my case and let you decide.

John Wayne and Gary Cooper

Bob Lynott is a crusty old retired weather forecaster in Portland, Oregon. I like Bob a lot, but there are people who don't. He's a maverick not afraid to speak his mind and rattle cages. By his own admission he's a gadfly. A gadfly, he says, "is a constructive critic, one who stimulates thinkers, and annoys the others." Bob wrote a book called *The Weather Tomorrow— Why Can't They Get It Right?* in which he took the National Weather Service to task. I think he was wrong in doing that, and I told him so; but he does make some interesting observations on weather forecasting in his book.

"Weather forecasting is not for the timid," he says. "Old forecasters never die. They become walking masses of scar tissue. They often are wounded before they learn to shoot accurately. Predicting the future is something like a shoot-out with John Wayne or Gary Cooper."

Bob goes on to warn against a currently popular term in the meteorological business, *nowcasting*. "The word nowcasting is a disguise to offer observing as an easier substitute for forecasting," he rails. "Nowcasting is the transvestite of the weather forecast world."

The sections that follow in this chapter don't offer weather forecasts in the traditional sense of the word. That is, no predictions of specific events for specific times are made; only a broad outlook is given. Despite that, I certainly feel like it's high noon or that Rooster Cogburn is lurking just around the corner. It would be a whole lot easier to nowcast for the 1990s . . . just let you know when the greenhouse effect hammers us. But once a forecaster, always a forecaster, I guess. As Bob says, we never die.

So I'll strap on my six-gun, but I'm really not looking for a shootout; I'm looking for that wolf we let in. We can't kill him, but at least we can try to throw a net over him and study him.

How Many Have Been Listening?

In my 1980 *Greenhouse Effect* (page 33) I suggested that the 1930s would provide a good climatic analog for the initial decade of CO_2-induced warming, a warming I expected to become noticeable about 2010. I thought we had time. I thought we'd start doing some things to ameliorate the greenhouse effect. We had a chance, and we had suggestions. But as I said earlier, "how many people have been listening? Only a handful, apparently."

Our appetite for fossil fuels has continued unabated since 1980. U.S. energy conservation efforts showed promise in the early 1980s, but the oil shortage became a glut, the economy took off, and our demand for energy followed suit. The development of nuclear power would have helped stem our use of fossil fuels, but the expansion of nuclear energy, if not dead in the United States, is certainly comatose, a victim more of emotion than of reason. The development of clean, renewable energy resources would have helped, too. But the Reagan Administration slashed support for the solar energy program by 80 percent. Another comatose victim.

To make matters worse, researchers have discovered that CO_2 isn't the only culprit in the greenhouse effect. There's methane, chlorofluorocarbons (CFCs), nitrous oxide, and ozone. And it turns out these "trace gases" are far more efficient contributors to the greenhouse effect than is carbon dioxide. So much so, in fact, that atmospheric chemists estimate the trace gases will shortly be just as important in the greenhouse effect as is CO_2. So much for hope.

And as if all of that weren't bad enough, consider the Amazon rain forests. In 1987 an estimated 120,000 square miles (310,880 square kilometers) of Amazon rain forest were destroyed by burning, logging, and flooding. The destruction contributed to 10 percent of worldwide CO_2 pollution. The good news, tongue-in-cheek, is that this will be a short-lived problem: shortly after the turn of the century, at the late-1980s rates of decimation, the Amazon rain forests will be gone.

What all of the preceding means (and all of the preceding is discussed in depth in chapter 6) is that we've apparently managed to hurry the greenhouse effect along. Given the warnings of the 1980s, and given recent global temperature trends, the analog of the 1930s can probably be applied to the latter part of the 1990s. We won't have to wait for 2010.

Why the 1930s? Why choose that decade as a model for our near climatic future? Prior to the 1980s, the 1930s boasted the pinnacle of recent global warming (see figure 1.1). The warming was particularly marked in the Northern Hemisphere, with specific dire consequences for North America. The rapid worldwide warming of the eighties and the warnings we've so far received (chapter 1) suggest that North America is again on the threshold of such consequences.

Why only on the threshold? Because Northern Hemispheric latitudes outside the tropics haven't quite yet attained the warmth of the thirties, indicating that—for North America—the most severe ramifications of near-term global warming are still to come.

But the analog of the thirties and recent trends aren't the only suggestions. There are other matters to be considered. But first, the 1930s.

Dust Bowls and Ice Packs

The decade of the 1930s was the warmest the world had seen in at least a century. The thirties were years of sweltering heat waves, searing drought, and heightened tropical storm frequency. Oceans warmed and ice packs retreated. Russian icebreakers in the Arctic cruised freely through regions now generally covered by unnavigable ice. And glaciers on Arctic islands rapidly dwindled.

In the United States a greater number of state records for high temperatures and dryness were set during the 1930s than in any other decade since the 1870s, when record keeping began (table 2.1). A superdrought turned the Midwest into a vast Dust Bowl, and people by the thousands fled the area.

Superdrought

The greatest disaster in American history attributable to meteorological factors was the superdrought of the 1930s. That awesome dry spell desiccated fifty million acres (20,250,000 hectares) of the Great Plains. Previous poor agricultural practices had bared the midwestern topsoil to the mercy of the elements, and the relentless winds that accompanied the drought swept away 350 million tons (317 million metric tons) of the richest soil in the world.

The first fingers of what was to become the great Dust Bowl drought crept into the central United States in late 1930. The dryness stretched out through early 1931 to grasp the northern and central Plains. Thereafter, every year through 1939 brought serious drought to some part of the Midwest.

In 1934 and 1936 the entire region from Texas to Canada was scourged. Dendrohydrologists (scientists who study climate records as revealed in tree growth-rings) have since determined that 1934 was the driest single year in the western United States since 1700. During that fateful year, fully 66 percent of the nation withered in severe-to-extreme drought. (By comparison, the figure for 1988's headline-making dry spell reached only 37 percent.)

The most seriously affected areas were in the south-central Plains. The contiguous parts of Colorado, New Mexico, Kansas, Oklahoma, and Texas formed the heart of the Dust Bowl. From 1934 through 1936, Dalhart, Texas, near the center of the region, averaged just 11.08 inches (281 mm) of precipitation per year, or 58 percent of normal. Between August 1932 and October 1940 the soil moisture index in western Kansas was continuously below normal, with thirty-eight months registering extreme drought.

The drought, coming simultaneously with the Great Depression, triggered a tremendous wave of emigration from the Plains, particularly from Oklahoma and adjacent states. John Steinbeck, in *The Grapes of Wrath*, vividly portrayed the movement of the "Okies" from the Plains to California and their plight upon arrival. He vividly portrayed, too, the wind and the dust: "The wind grew stronger. The rain crust broke and

Table 2.1. State Temperature and Precipitation Extremes
Occurring Each Decade since 1870

State Record Maximum Temperature*		State Record Minimum Temperature*	
Decade	Number of State Records	Decade	Number of State Records
1870–1879	0	1870–1879	0
1880–1889	1	1880–1889	0
1890–1899	2	1890–1899	6
1900–1909	2	1900–1909	6
1910–1919	5	1910–1919	4
1920–1929	3	1920–1929	3
1930–1939	27	1930–1939	11
1940–1949	1	1940–1949	4
1950–1959	6	1950–1959	4
1960–1969	1	1960–1969	5
1970–1979	2	1970–1979	4
1980–1989	1	1980–1989	4

the dust lifted up out of the fields and drove gray plumes into the air like sluggish smoke. The corn threshed the wind and made a dry, rushing sound. The finest dust did not settle back to earth now, but disappeared into the darkening sky."

Black Blizzards

And darkening skies there were. The first of a series of monster dust storms, or "black blizzards," struck in November 1933. Vast quantities of dust were swept thousands of feet into the air from Montana to the western Ohio Valley. Dust reached the Atlantic seaboard from Georgia northward; "black rain" fell in New York State, and "brown snow" in Vermont.

Fearsome dust storms continued to ravage the Midwest until 1939. In 1934 and 1935 skies were obscured over the Ohio Valley and Great Lakes. Sunshine was dimmed in the eastern United States, and dust particles sifted down on ships in the Atlantic Ocean.

A Texas youngster during the 1930s remembers the black blizzards:

	*State Record Maximum Annual Rainfall**			*State Record Minimum Annual Rainfall**	
Decade	*Number of State Records*		*Decade*	*Number of State Records*	
1870–1879	2		1870–1879	0	
1880–1889	5		1880–1889	0	
1890–1899	2		1890–1899	2	
1900–1909	2		1900–1909	0	
1910–1919	1		1910–1919	1	
1920–1929	0		1920–1929	1	
1930–1939	4		1930–1939	18	
1940–1949	8		1940–1949	5	
1950–1959	11		1950–1959	12	
1960–1969	7		1960–1969	11	
1970–1979	5		1970–1979	0	
1980–1989	0		1980–1989	0	

Other records were set in 1845, 1851, 1853, and 1869

Another record was set in 1826

*Includes Washington, D.C.

"These storms were like rolling black smoke. We had to keep the lights on all day. We went to school with the headlights on and with dust masks on. I saw a woman who thought the world was coming to an end. She dropped down on her knees in the middle of Main Street in Amarillo and prayed out loud: 'Dear Lord! Please give them another chance.'"

In the cauldron that was the Dust Bowl, livestock died of suffocation and starvation; crops that survived were withered and stunted. Huge drifts of loose soil surrounded farms and blocked highways and railroads. The blowing dust blasted paint off houses and automobiles. Hundreds of people died of respiratory ailments, and thousands fled the area, simply abandoning their ranches to the wind and dust.

The black blizzards left profound physiological and psychological impacts on a generation of midwesterners. But drought and dust weren't the only meteorological adversaries people faced in the thirties. There was the heat. Penetrating, incinerating, unending.

Superheat

Between 1930 and 1936 nearly 15,000 people in the United States were killed by scorching summer heat. In 1936 alone, 4,768 lives were claimed by soaring temperatures and burning sunshine. In a typical year in the United States about 175 people die from excessive summer heat and sun; in the heat wave of 1983, 220 people lost their lives. Of course, air conditioning was not in widespread use in the thirties, but that doesn't make the death toll any less appalling.

The first of the great 1930s heat waves appeared in the midsection of the nation in July 1930. The mercury soared to state record levels in Tennessee, Kentucky, Mississippi. Perryville, Tennessee, tagged 113° F (45° C); Greensburg, Kentucky, registered 114° F (46° C); and Holly Springs, Mississippi, sizzled in 115° F (46° C) heat. In the East, Delaware tallied a state record 110° F (43° C), and Washington, D.C., hit a record-breaking 106° F (41° C).

The heat persisted into August. In Iowa, August 3 became the hottest single day in history. The average maximum temperature throughout the state that broiling Sunday was 106.4° F (41.3° C). It was but a harbinger of things to come.

Record hot spells continued in 1931. In April, Pahala, Hawaii, established the Hawaiian mark for heat with a reading of 100° F (38° C). Abnormally high temperatures in late June and early July in the Midwest killed thousands of field horses in Iowa. In Florida, Monticello racked up that state's highest reading ever—109° F (43° C)—on June 29. By the time the year was done, figures showed 1931 to have been one of the warmest years on record. Every section of the country had had above-normal temperatures.

The intense heat waves relented a bit during the next two years, 1932 and 1933. But the winter of 1931–32 was the mildest ever around the Great Lakes and in parts of the Northeast; in Baltimore it was the warmest winter in 115 years.

The oppressiveness returned in earnest in 1934. In Kansas and Iowa the summer was even hotter than in 1930. It was described as the worst crop season in history; the corn was virtually wiped out. Keokuk, Iowa, fried in 118° F (48° C) heat, and Gallipolis, Ohio, suffered a 113° F (45° C) reading. Both temperatures are state records. Other all-time state marks were established that summer in Idaho (Orofino, 118° F [48° C]) and New Mexico (Orogrande, 116° F [47° C]). In Cincinnati, Ohio, it was the hottest summer on record, with the thermometer topping off at 109° F (43° C) in July.

The heat backed off again in 1935, but in a figurative sense it was the calm before the firestorm. The heat during the torrid summer of 1936 was to become legendary.

In July and August 1936, the heat records of 1934, which had topped those of 1930, were themselves exceeded. The heat was phenomenal. C.D. Reed, head of the Iowa Weather and Crop Bureau, made an extensive study of past heat conditions in his state. He drew these conclusions about July 1936: "Comparing the mean temperatures at individual stations in all Julys back to the beginning of records in 1819 with the July 1936 mean temperatures . . . , there is ample margin in favor of 1936 to take care of all possible differences due to location and exposure of instruments and methods of observation, and still leave July 1936 well in the lead as the hottest July in 117 years."

The prostrating Iowa heat reached its pinnacle on the afternoon of July 14. The average maximum for 113 stations across the state was an astounding 108.7° F (42.6° C)! This topped the nearest challenger, that hot August Sunday of 1930, by more than 2° F (1.1° C).

Sleeping on Sidewalks

Kansas City, Missouri, the largest city that borders the Plains, also put records in the book that sweltering summer. The mercury soared to 100° F (38° C) or higher on fifty-three days! The first time came on June 15, with the heat peaking on August 14 at 113° F (45° C), the city's hottest ever. Residents that summer slept outdoors at night—in yards, on roofs, on sidewalks—searching for small degrees of comfort.

To provide an opportunity to compare the blast furnace of 1936 with recent periods of intense heat in Kansas City, table 2.2 lists the July and August daily maximum temperatures for 1936, 1980, and 1983. Although maxima for July 1980 averaged slightly higher than those of July 1936, neither 1980 nor 1983 came close to matching 1936 for the number of times the mercury soared to 100° F (38° C) or more.

The July–August 1936 period was the steamiest on record not only in Kansas City but in other cities in the Midwest and Great Plains as well. From Oklahoma City northward to Bismarck, North Dakota; from Omaha, Nebraska, eastward to Cincinnati, Ohio, new records from summer heat were set that blazing year.

Most remarkable of all, state records of 121° F (49° C) were established in North Dakota and Kansas, and of 120° F (49° C) in South Dakota,

Table 2.2. Downtown Kansas City High Temperatures
during Three Record Hot Summers: 1983, 1980, 1936

	1983	1980	1936		1983	1980	1936
July 1	94	109	93	August 1	98	108	90
July 2	88	87	93	August 2	100	97	93
July 3	93	94	103	August 3	100	101	97
July 4	82	103	108	August 4	98	99	101
July 5	82	103	103	August 5	99	96	79
July 6	86	105	99	August 6	88	95	80
July 7	86	106	100	August 7	94	100	79
July 8	90	105	100	August 8	94	100	98
July 9	90	109	101	August 9	97	101	110
July 10	92	109	104	August 10	99	101	102
July 11	95	108	104	August 11	92	87	99
July 12	96	103	105	August 12	87	94	109
July 13	92	109	105	August 13	92	99	110
July 14	91	109	109	August 14	86	87	113
July 15	88	109	107	August 15	96	84	110
July 16	92	102	106	August 16	103	89	104
July 17	91	104	105	August 17	104	92	108
July 18	98	108	104	August 18	100	95	108
July 19	99	108	102	August 19	100	98	107
July 20	99	106	93	August 20	94	98	103
July 21	101	92	92	August 21	100	89	104
July 22	101	88	102	August 22	90	91	104
July 23	102	90	102	August 23	87	91	104
July 24	96	97	109	August 24	99	95	109
July 25	86	97	108	August 25	102	98	107
July 26	89	89	106	August 26	103	98	106
July 27	102	88	108	August 27	105	91	104
July 28	103	102	95	August 28	93	91	91
July 29	98	102	93	August 29	95	94	86
July 30	95	109	84	August 30	85	94	90
July 31	98	103	88	August 31	91	97	94

Texas, Arkansas, and Oklahoma. *These are the only cases of readings of 120°
F (49° C) or higher ever being reached outside the southwestern desert triangle
of California, Nevada, and Arizona.* (There was one doubtful exception in

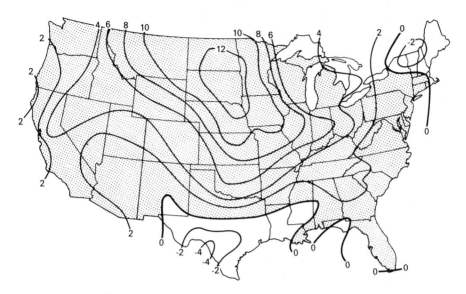

FIGURE 2.1 U.S. temperature anomalies, July 1936.

Oklahoma in 1943 when a temperature of 120° F [49° C] was reported, but no other station on the date in question came within 8° F [4.4° C] of 120° F, and the figure is thus suspect.) Elsewhere in the Midwest, Minden, Nebraska, soared to 118° F (48° C) during the 1936 hot spell, while locations in Minnesota and Wisconsin hit 114° F (46° C), all state records.

The superheat that summer moved eastward and southward, too, but not for extended visits. Still, the hottest readings ever were recorded in Indiana (Collegeville, 116° F [47° C]), Louisiana (Plain Dealing, 114° F [46° C]), Maryland (Cumberland and Frederick, 109° F [43° C]), Michigan (Mio, 112° F [44° C]), New Jersey (Runyon, 110° F [43° C]), Pennsylvania (Phoenixville, 111° F [43° C]), and West Virginia (Martinsburg, 112° F [44° C]).

Figure 2.1 shows the average temperature deviations over the United States during the great hot month of July 1936. For comparison, figure 2.2 graphs the temperature departures during the more recent heat wave of June 1988. The heat anomalies of 1988 were just as intense of those of 1936; but overall, 1988's above-normal readings were confined to a smaller area. More important, the anomalies of 1988 occurred in June rather than July. June mean temperatures are lower than those of July; thus a departure of plus 10° in June implies weather not quite so hot as a 10° anomaly in July.

By the late 1930s the monster heat waves began to disappear, but not

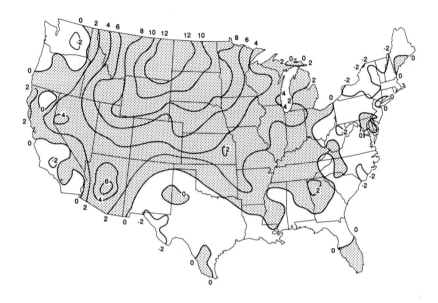

Figure 2.2 U. S. temperature anomalies, June 1988.

Figure 2.3 State high temperature records established during the 1930s.

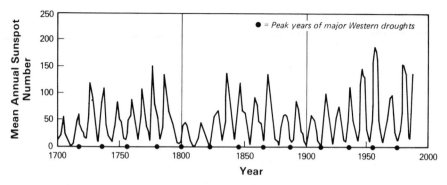

F<small>IGURE</small> 2.4 Annual smoothed sunspot numbers with peak years of major western droughts.

before leaving a state record of 117° F (47° C), in Medicine Lake, Montana, in 1937. Figure 2.3 summarizes state high temperature records set during the 1930s. Heat equal to the widespread blistering extremes of the 1930s has not since returned to the United States, but it will as we inexorably alter our climate with CO_2.

The wolf is in the door. But there is more behind his breath than the trend of recent global warming and comparisons to the climate of the thirties. There is a frightening cycle of major drought in the western United States that is due to return in the 1990s, independent of any greenhouse effect.

Droughts and Sunspots

On a broad time scale the occurrence of major droughts in the United States west of the Mississippi River appears to follow a fairly regular cycle of twenty to twenty-two years and to closely match the "double sunspot cycle"—that is, two eleven-year cycles of sunspot activity.

Sunspots are relatively small, dark areas that appear on the sun's surface. In reality they are great magnetic storms that swirl through the hot, gaseous photosphere of the sun. Sunspot activity progresses through an irregular cycle of roughly eleven years during which the number of spots moves from a minimum to a maximum and back to a minimum. The number of actual sunspots may be as low as zero or as high as 200 on an average annual basis. And the cycle may vary in length from ten to fifteen years.

Figure 2.4 shows the sunspot cycles back to 1700. Major droughts, also

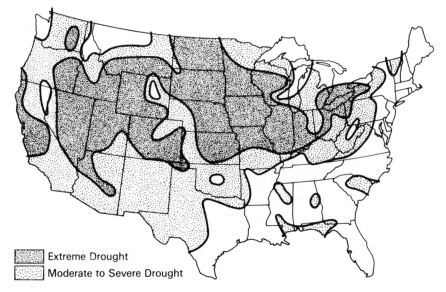

Extreme Drought
Moderate to Severe Drought

FIGURE 2.5 A "snapshot" of the 1930s drought at its peak in July 1934.

indicated on the figure, in the western United States have occurred in conjunction with every other sunspot minimum back to 1700. This drought-sunspot link, at least in a chronological sense, is particularly strong on the Great Plains, where tree growth-ring analyses provide evidence for the period prior to written history.

Large-scale settlement of the Plains began after the Civil War. From the mid-1860s to the mid-1880s plentiful rainfall in the region encouraged the immigration of farmers and ranchers. But precipitation decreased significantly after 1886, and settlers in western Kansas and Nebraska began a massive exodus. As an historian of the period wrote, "Fully half the people of western Kansas left the country between 1888 and 1892." And things got worse after that. Extreme drought seared the region in 1893 and 1894, culminating in total crop failure on the Plains in 1894. The sunspot cycle had reached a minimum in 1889.

The nadir of the double sunspot cycle was reached again in 1913, and serious but short-lived droughts stalked the prairies in 1910, 1911, and 1913. Then came the great Dust Bowl and the black blizzards of the 1930s. The sunspot minimum: 1934. Figure 2.5 provides a "snapshot" of the 1930s drought at its peak in July 1934. The snapshot is based on the Palmer Index, a measure of meteorological drought calculated from temperature and precipitation data as recorded by instruments.

The next double sunspot cycle bottomed out in 1953. And again there

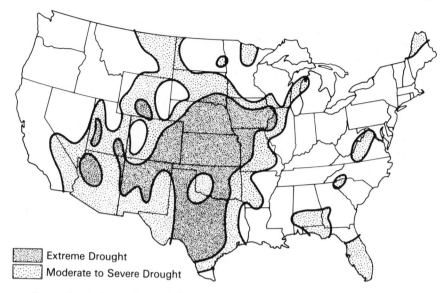

FIGURE 2.6 A "snapshot" of the 1950s drought near its peak in July 1956.

was a major drought on the Plains, this one shorter, more confined, and centered a bit farther south than its 1930s predecessor. But in Texas it was worse than the Dust Bowl dry spell. For almost five years, from mid-1952 to early 1957, most of the state did not get what old-timers call "a public

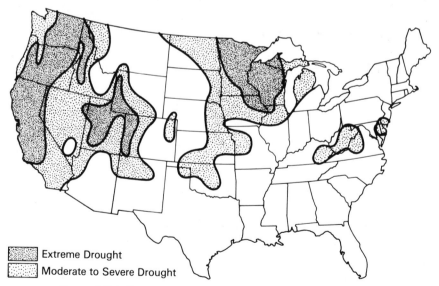

FIGURE 2.7 A "snapshot" of the 1970s drought in April 1977.

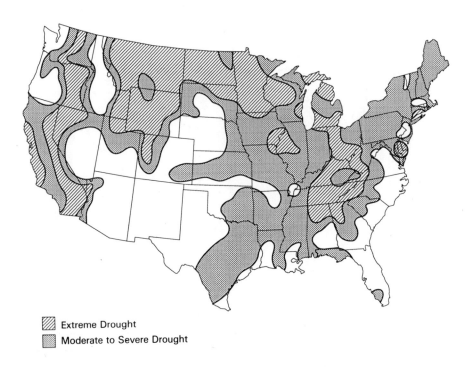

Extreme Drought
Moderate to Severe Drought

FIGURE 2.8 A "snapshot" of the 1988 drought in July.

rain." Major rivers could be jumped on foot, and the bottom of giant reservoirs lay naked and cracked. In Oklahoma, farm families were fleeing the desiccated land at the rate of 4,000 per day. Figure 2.6 is a Palmer Index picture of drought conditions in July 1956.

The drought of the 1970s, coming near the sunspot minimum of 1976, spared much of the Plains. But it hit hard in the Far West and Rocky Mountain states and attacked Minnesota and Wisconsin. In the West, snowpacks disappeared, stream flows dwindled, and urban water rationing became a reality. Figure 2.7 is a snapshot of the drought in April 1977 as the western United States came out of a virtually snowless winter. As a point of comparison, figure 2.8 provides a look at the 1988 drought near its peak in July.

Next!

The drought–sunspot cycle connection, chronologically, is statistically strong. Researchers, however, have not been able to establish what is known as cause and effect; that is, they don't know why every other sunspot minimum should cause a drought. Still, the time link is there. And that brings us back to the 1990s.

The most recent sunspot minimum occurred in 1986 but was not the part of the double sunspot cycle tied to drought. That part is due in the 1990s!

To find out just when, I talked with Patrick McIntosh at the Space Environmental Services Center in Boulder, Colorado. McIntosh says that with one exception, the sunspot cycles have been quite regular this century. Thus he feels confident in predicting that the next sunspot minimum will come in 1997.

So another circumstantial clue can be thrown on our 1990s pile suggesting intense heat and major drought. The drought-sunspot con-junction doesn't tell us *exactly* when or where a major drought might strike in the 1990s, though . . . so back to speculation.

Timing is most difficult. Over the past 100 years, sunspot cycle–linked droughts appear to have been initiated anywhere from one to four years before the time of sunspot minimum. Thus, keying on the 1997 date, we might look for the first signs of major drought as early as 1993 or as late as 1996. Historically, some sunspot cycle droughts haven't peaked until slightly after the sunspot minimum (see figure 2.4). So the genesis of extreme dryness could be delayed even beyond 1996. Whatever the timing, I believe strongly that severe to extreme drought will strike before the end of the decade.

Where? Here we can turn to the warm-world analog of the 1930s for help. The atmospheric circulation patterns of the 1990s should more closely resemble those of the 1930s than of any other recent period. Thus, although much of the nation could be gripped by drought in the 1990s, the 1930s parallels suggest that the northern and central Plains and southern Canada will be the focus of the attack (see figure 2.5). None of this is very precise, of course, but it's the best that can be done. Remember, it's speculation.

And surprisingly, in a broad sense, I have an ally. James Hansen, coming from a different direction—the world of computer models—has caught a glimpse of the wolf (though he wouldn't call it that) in his work for the National Aeronautics and Space Administration (NASA). At a meeting in August 1988 he told the National Governor's Association Committee on Energy and Environment, "My personal opinion, based on available information and models, is that we are likely to have more severe and widespread summer droughts in the next decade or two than was the case for the period . . . 1950–1980."

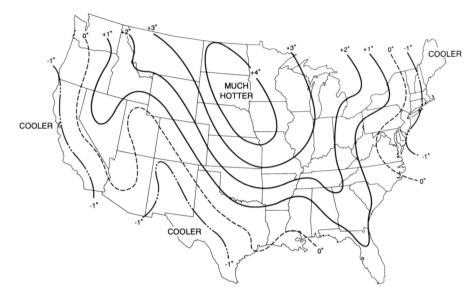

FIGURE 2.9 The mean July temperature differences between the 1930s and now.

How Hot? How Dry?

How intense might be the heat and drought foreseen for the nineties? Again, only speculation. Certainly the analog of the thirties should prevail.

First of all, the heat. The biggest differences in mean monthly temperatures between the 1930s and now were in the summer months. Julys in particular were noticeably hotter in the 1930s, averaging 4° F (2.2° C) or more warmer over much of North Dakota, South Dakota, Nebraska, and Iowa. Figure 2.9 provides a graphic look at that.

Individual Julys during the Dust Bowl decade had mean temperature departures considerably above 4° F (2° C), of course. For instance, at Lincoln, Nebraska, July 1936 averaged almost 10° F (6° C) above the modern mean! An *average* difference of 4° F persisting for ten years is really quite remarkable when you consider that at Lincoln, for example, the hottest single Julys since the 1930s have averaged only about 4° or 5° F (2.5° C) above the current normal. Lincoln's hottest day ever came in July 1936: 115° F (46° C). Remember, too, that the 1930s saw temperatures skyrocket to 120° F (49° C) or more as far north as the Dakotas.

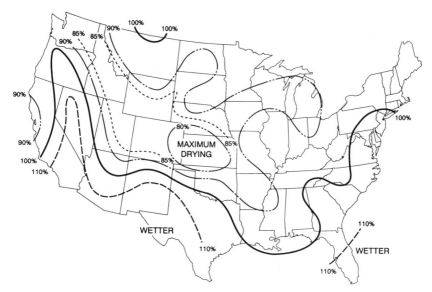

FIGURE 2.10 The pattern of average yearly precipitation in the 1930s as a function (percentage) of current normals.

Back to Hansen's NASA computer models for a moment. (So such things don't appear to be total black magic, computer models and climatic analogs are discussed in chapter 7.) One of the greenhouse-effect climate simulations for the 1990s run by the NASA group produced summertime (June-through-August) temperature deviations quite similar to the July temperature departures shown in figure 2.9. The NASA output indicated that under conditions of unabated CO_2 and trace gas output (and no volcanoes), summers of the 1990s would average 2° to 4° F (1° to 2° C) warmer from the Pacific Northwest eastward to the northern Plains, then southeastward to encompass a broad area of the Midwest and Deep South. Hansen has pointed out that according to NASA's model, the hot summer of '88 would be a "normal" one in about ten years.

The model also showed summertime *cooling* in the Southwest, roughly similar to that suggested by figure 2.9. And this highlights an important point. Under greenhouse-effect conditions, not all regions of the United States or of the world would grow immediately warmer or drier. There would be many regional variations, and these are discussed in chapter 5.

It is also important to note that even a trend toward warmer, drier conditions would not occur smoothly. Natural weather variations would continue to dictate occasional cooler, wetter seasons, including—in the case of the northern United States—some harshly cold winters. Such fluctuations are also discussed in chapter 5.

Now the drought. Maximum drying during the thirties, relative to now, occurred in the center of the nation, particularly from eastern Colorado through much of Kansas. There, for an entire decade, average annual precipitation was at least 20 percent below current normals. Figure 2.10 shows the pattern of average yearly precipitation over the United States during the 1930s as a function (percentage) of current means.

Even more striking—and potentially more devastating—than the pattern of annual precipitation changes was the scope of July drought across the Great Plains. In most of Kansas and Oklahoma, July rainfall during the 1930s averaged less than half the modern normal!

The Great Drought of the 1990s

So there it is. Precursors. Global temperature trends. Climatic analogs. Sunspot cycles. Computer models. They're all part of the net we tried to throw over that wolf. We got the net on him, but he didn't hold still. We need a better net. But we won't get one soon enough. We'll have to settle for a rough examination of a thrashing, cantankerous beast who won't give us any more time. He has given us an unmistakable warning, though.

The Great Drought of the 1990s? It isn't fantasy. It's something that's happened before, and we'd be foolish to think it won't happen again (though not *exactly* as it did in the thirties, since analogs serve only as guides, not as precise predictors). There's a quote to keep in mind, and I don't know whether it was first uttered by a climatologist or an historian. At any rate, it applies to our current situation: "What has happened, can happen." We'll be lucky if that's all that happens.

3/ Supercane

800 Millibars

Shortly after Hurricane Gilbert, the "granddaddy of all storms," thundered across Jamaica in September 1988, a U.S. Air Force reconnaissance craft penetrated the storm's eye. In the eye, the crew aboard the aircraft recorded the lowest atmospheric pressure ever measured at sea level in the Western Hemisphere: 888 millibars. A millibar is a measure of air pressure used by meteorologists. For those of us more comfortable with home barometers, a pressure of 888 millibars converts to 26.22 inches.

Perspective

Ever since I was a kid, I've been fascinated by weather. I got my first barometer—which after forty years still sits on my dresser at home—when I was ten years old. I remember how excited I was the first time I saw a reading below 29.00 inches: 28.80 inches, to be more precise, or 975 millibars. Readings of that magnitude are about what a typical strong winter storm in the Pacific Northwest or a nor'easter in New England might be expected to produce. (The average atmospheric pressure at sea level is 1013.2 millibars, or 29.92 inches.)

The lowest sea-level pressure ever observed was in a typhoon in the Pacific Ocean, about halfway between Guam and the Philippines. (Typhoons are the same as hurricanes; they just go by a different name west of the International Dateline in the Pacific.) In October 1979 the air force again had the dubious honor of being vectored into an unbelievably intense tropical cyclone. In the center of Typhoon Tip, a crew from the 54th Weather Reconnaissance Squadron recorded a pressure of 870 millibars (25.69 inches). And as they careened through the maelstrom, they estimated sustained winds at 160 knots (184 mph), with gusts to 195 knots (224 mph). Now that's a storm! It might, however, be a wimp compared with what's in our future.

Supercanes

Kerry Emanuel of the Massachusetts Institute of Technology ran some calculations on what storms of the future might be like. Hurricanes and typhoons gather much of their energy from the warmth of tropical seas; the warmer the surface water, the greater the potential strength of a tropical cyclone. So Emanuel looked at how much warmer tropical oceans might get in response to the atmospheric CO_2 concentration reaching twice its preindustrial level. At the earliest, this warming probably would not be attained until the middle of next century. Despite that, the conclusions Emanuel drew were startling and certainly have ramifications for us in the near term.

Admittedly, a lot is unknown about how rapidly ocean temperatures will increase in response to the greenhouse effect. Even if they fail to rise as quickly as anticipated, the trends that Emanuel's calculations suggest are frightening.

For instance, in the northwest Gulf of Mexico, water temperatures several decades from now may be warm enough to produce the potential of a hurricane's central pressure plunging to 800 millibars (23.62 inches). Remember, Granddaddy Gilbert's lowest reading was 888 millibars (26.22 inches). An 800-millibar "supercane" would produce sustained winds of 205 knots, or 235 mph!

Of course, only a very few hurricanes would attain that awesome severity, and while Emanuel didn't explicitly study the *average* strength of hurricanes under a warmer-ocean scenario, he pointed out that "one intuitively expects that average intensity increases with maximum intensity."

And more storms, too? Not necessarily, according to Emanuel. The conditions that breed hurricanes are not the same ones that strengthen them, so it isn't obvious at this point there would be a substantially greater number of tropical cyclones in a greenhouse world . . . just that there would be a potential for some immensely more powerful ones. On the other hand, there are scientists who believe the changing atmospheric circulation patterns brought on by the greenhouse effect will indeed be conducive to the formation of more tropical storms and hurricanes.

Gilbert, Allen, and Labor Day

So all of this brings us back to the 1980s and the 1930s, the two warmest decades on record in the Northern Hemisphere. These two decades produced the three most intense hurricanes ever observed in the Atlantic–Caribbean–Gulf of Mexico region. The 1980s brought Gilbert (1988) and Allen (1980). (Allen was mentioned in chapter 1.) In Allen the lowest pressure measured was 899 millibars, or 26.55 inches. At the time Allen was the second strongest hurricane ever seen, surpassed only by a storm in 1935, the infamous Labor Day Hurricane. In that storm, an observer on Matecumbe Key, near the middle of the Florida Keys, reported a barometer reading of 26.35 inches. (Millibars weren't yet in use, but the reading converts to 892 millibars.)

Any hurricane with a central pressure less than 920 millibars (27.17 inches) is classed as a category five hurricane on an intensity scale of one to five. The damage levied by a category five storm is described as "catastrophic." Both Gilbert and Allen weakened to category three hurricanes before making landfall in or near the United States, but the Keys Labor Storm didn't. It churned through the Keys as a full-blown five. It ripped the chain of islands with wind blasts in excess of 175 knots (200 mph) and storm tides fifteen to twenty feet (4.5 to 6 m) above normal. Wave wash marks on trees indicated water thirty feet (9m) above sea level! A World War I veterans camp was inundated, and the railroad over the Keys was swept away. An eleven-car rescue train dispatched to Lower Matecumbe Key was virtually washed off the tracks. Only the locomotive and tender survived. The death toll from the storm reached 408. Most victims were engulfed by a wall of water, but some were literally sandblasted to death; they were found with eroded skin and no clothes save for belt and shoes.

Certainly that thunderous Labor Day storm can't be attributed to the greenhouse effect. For that matter, Gilbert and Allen probably can't be either. Probably. But one has to reflect on the fact that all three storms occurred during periods of significant global warming and that the current warming is likely to continue, perhaps at an accelerating rate. As the atmosphere grows warmer and warmer, the oceans follow suit, although they are much slower to respond to temperature changes than are land masses. (Curiously, a recent National Oceanic and Atmospheric Administration study concluded that ocean temperatures between 1982 and 1988 rose twice as much as previously thought. Chapter 7 talks more about this study.)

At some point, probably by the middle of next century, sea waters may well be warm enough to support 800-millibar superstorms. That's several decades away, however, so why should we worry now? The point is, of course, that the potential for such storms does not develop in a discontinuous manner. That is, we won't suddenly jump from 888-millibar Gilberts to 800-millibar supercanes fifty years from now. The process will be gradual. Each year as the temperatures of sea surfaces gradually elevate, the potential for blockbuster hurricanes will, too. One has to suspect that during the 1990s at least the *potential* for, if not the reality of, a hurricane exceeding the great Gilbert's fury will exist. And next time we might not be so lucky. Next time a category five storm might not weaken to a category three before it slashes across a U.S. coastline.

The Long Island Express and Other Legacies of the Thirties

Since the 1930s are being proposed as an analog for the 1990s, it seems appropriate to review the tropical storm and hurricane activity that occurred during the 1930s.

Although the decade of the thirties did not produce an annual average number of tropical storms and hurricanes that is any greater than the current average, it should be kept in mind that the count was made without the aid of aircraft and satellite surveillance. There may well have been storms that went undetected. The thirties, however, did produce the two most active Atlantic–Caribbean–Gulf of Mexico hurricane seasons on record: 1933 and 1936. Twenty-one tropical storms and hurricanes blossomed in 1933 and sixteen in 1936. The current mean is ten, and the most active year in the last three decades was 1990, with fourteen named storms (1969 saw thirteen).

As another point of comparison, in the period 1930–39, eight hurricanes classed as category three ("extensive" damage) or greater rammed into U.S. shores. Although this is not a record number, it is a larger tally than recent decades have produced. During 1960–69, six hurricanes reaching or exceeding category three slammed into the United States. The 1970–79 period produced four such storms, while the 1980s brought five.

Most of the category three or stronger hurricanes that crossed U.S. coastlines during the 1930s smashed into either Gulf Coast states or Florida. One storm battered New England, another North Carolina. The pattern of more major storms hammering Florida and the Gulf Coast than moving up the East Coast is consistent with the type of weather regimes that prevail during heat and drought in the midsection of the country. During such periods, upper-level (steering) winds blow clockwise around a large high-pressure cell entrenched over the center of the United States. Steering winds are thus northerly (i.e., blowing from the north) over the eastern seaboard much of the time, discouraging hurricanes from moving northward. Over Florida and the Gulf Coast, steering winds are often easterly or southeasterly, driving tropical storms and hurricanes from the Atlantic and Gulf of Mexico into the southern states.

Ironically, though, the most devastating storm of the thirties hit New England. The Great New England Hurricane of 1938 (technically a category three storm) was not nearly so intense as the Keys Labor Day storm, but it killed more people and did more damage.

The Express

In 1938 it had been almost seventy years since New England had been assailed by a severe hurricane. Even then the damage had been confined to a relatively small area of southeastern New England. So the news that on September 21, 1938, a severe hurricane was spotted off Cape Hatteras, North Carolina, was met with relative indifference by both the public and the U.S. Weather Bureau. Everyone knew that hurricanes veered out to sea before they could slam into New England.

Wrong. By nightfall, close to 700 people were dead in New England and on Long Island, New York. Property damage was staggering: 387 million dollars at depressed 1938 prices. New London, Connecticut, lay in ruins; Providence, Rhode Island, was under thirteen feet (4m) of water; and Milton, Massachusetts (near Boston) had been ripped by winds up to 162 knots (186 mph). The Great New England Hurricane had charged northward without swerving. It had raced from Cape Hatteras to Long Island at speeds up to sixty knots (70 mph), and later became known as the "Long Island Express." The Express had thundered into the Northeast with no warning and at high tide. A more deadly combination of circumstances could not have been imagined.

Elsewhere along the East Coast, North Carolina and Virginia were the targets of several hurricanes during the 1930s, although only one storm struck the area as a major cyclone. The storms were unique in that they swirled into the Cape Hatteras and Norfolk, Virginia, areas directly from the southeast on tracks not since duplicated. The first of the storms struck on August 23, 1933, with the eye sweeping directly over Hatteras, then moving up the James River in Virginia. Winds reached sixty-one knots (70 mph) at the Norfolk airport and seventy-seven knots (88 mph) at the Naval Air Station. Ten-foot (3-m) tides surged through both Norfolk and Hampton, Virginia. High-water marks hit twelve feet (3.7 m) in the upper reaches of the Potomac River. Seashore resorts sustained extensive damage, but the death count was relatively low: eighteen.

Three weeks later a second tropical storm charged into Cape Hatteras from the southeast but hooked away sharply to the right before it could attack Hampton Roads. Tides ran about a foot (0.3 m) lower than in the previous storm, with the Outer Banks of North Carolina catching the cyclone's full fury. Winds at Hatteras were clocked at sixty-five knots (75 mph), and thirteen inches (330 mm) of rain pounded down. Twenty-two lives were lost, mostly in New Bern, North Carolina, where water two to four feet (0.6 to 1.2 m) deep surged through the streets.

A third hurricane lumbered into Cape Hatteras from a southeasterly direction on September 18, 1936. But this powerful storm, with winds in excess of 117 knots (135 mph) over the open ocean, relaxed a bit as it brushed the Outer Banks and began to curve away from the mainland. Still, tides five to six feet (1.5 to 1.8 m) above normal roiled through Hampton Roads.

The year 1933 saw the busiest hurricane season ever. Florida alone was bombarded by four storms that year. On September 3 and 4, Jupiter Inlet, on Florida's east coast, reeled under the most powerful blow. Winds of 109 knots (125 mph) hammered the area, although residents found themselves in the relative calm of the storm's eye for a full forty minutes.

Texas also caught the brunt of four storms in 1933. On the same date that Jupiter Inlet was being battered, winds of seventy knots (80 mph) howled through Brownsville, Texas. Tides built to twelve to fifteen feet (3.7 to 4.6 m), and forty people perished. Other strong (category three) hurricanes during the 1930s clobbered the upper Texas coast (1932), Louisiana (1934), and the Florida panhandle (1936).

And it was Florida, more than any other state, that felt the impact of the often active hurricane seasons of the thirties. That decade delivered more tropical storms and hurricanes to the Sunshine State than any other decade this century. An average of almost two storms per year swept into the state, and one of the storms, the Keys Labor Day Hurricane, still is the second most intense on record in the Western Hemisphere.

A Hurricane Watch Has Been Issued for the California Coast . . .

While the 1930s produced some unusual hurricane activity in the
Atlantic, there were some surprising events in the eastern Pacific, too.
Tropical storms and hurricanes forming off the Pacific coast of Mexico
typically move westward into the open ocean. Or, if they do trek
northward, they weaken significantly as they approach the United States.
In 1939, however, an anomalous number of full-fledged tropical
cyclones—five—in response to warmer ocean waters and altered atmospheric circulation patterns, crossed the North American coast from Baja
California northward.

Two storms passed less than 200 miles (322 km) south of San Diego,
while a third one, on September 25, blasted inland near Los Angeles.
Unprecedented gales and torrential rains caused damage in excess of two
million dollars and took forty-five lives. Los Angeles was drenched with
5.5 inches (140 mm) of rain, while Mount Wilson was deluged with
thirteen inches (330 mm).

So the 1930s not only fostered increased hurricane and tropical storm
activity in Florida and along the Gulf Coast but also delivered a most
unusual tropical cyclone to southern California.

There is reason to be concerned about what the 1990s hold in store.

Madness

In truth, even without the threat of increased tropical storm activity,
unusual tracks, and supercanes, there is reason to be concerned. And one
of the concerns has already become manifest.

In September 1989, Hurricane Hugo, a category four storm, slashed
into the South Carolina coast. From Charleston to Myrtle Beach, Hugo's
powerful winds and swirling storm tides swept away buildings, bridges,
and boats. The damage was record-breaking, approaching eight billion
dollars. In the initial draft of this book—prepared prior to Hugo—I had
been particularly apprehensive about the hurricane threat to the Carolinas. I had written:

> Consider, for example, the Carolina coast, which has been spared the
> brunt of a major hurricane for three decades. This makes that area
> particularly vulnerable, since a generation without the memory of a severe

hurricane has settled along the coast there. Much of the Carolina coast was swept clean by Hurricane Hazel in 1954, and battered again by Hurricane Gracie in 1959. But since that time land development in many coastal resort areas has exploded. For instance, in Myrtle Beach, South Carolina, development has taken place so rapidly since the late 1960s that homes and motels have been constructed on sand spits barely above sea level. Madness. Hazel generated storm tides eighteen feet (5.5 m) above sea level!

There was nothing especially prescient about my apprehension . . . the South Carolina coast had just been lucky for thirty years. Unfortunately, there are other recently developed areas along the Carolina coasts still under the gun: Long Beach Island (near Wilmington, North Carolina) and Hilton Head, Seabrook, and Kiawah Islands (off the South Carolina coast). These are barrier islands all previously denuded by hurricanes. Their luck will run out sooner or later, too. But Carolina coastal islands aren't the only ones in trouble.

The worst single natural disaster in American history was triggered by a hurricane that crashed into Galveston Island, Texas, in 1900. A high tide twenty feet (6 m) above normal, pushed by furious winds, churned through Galveston city. With inadequate warning, thousands of people were trapped in the city after the one bridge to the mainland had been washed away. The consequence was numbingly tragic: 6,000 people lost their lives within a matter of hours.

Today a seventeen-foot (5-m) seawall protects Galveston. But inexplicably, condominiums have been erected on a sandbar in front of the seawall. Madness. On South Padre Island, along the extreme southern Texas coast, numerous sand dunes have been leveled to make way for high-rise hotels and condos. And where the dunes have not been leveled, buildings have been merely placed on top of or in front of them. Ironically, the only natural defense that barrier islands such as South Padre have against the ravages of hurricanes is the natural dune system. Some officials have predicted that not a structure will be left standing on South Padre after the next major hurricane thunders in.

Of even greater concern, of course, is the potential for a staggering loss of life. In 1988 Robert Sheets, director of the National Hurricane Center (NHC), said, "It is more clear than ever that hundreds of lives could be lost if a major hurricane sweeps across a large population center. . . . The recent completion of hurricane coastal flood models for the Atlantic and Gulf coasts reveals that many more people than originally thought must be evacuated under certain hurricane conditions." Over forty-three million people currently live in 175 coastal counties from Texas to Maine.

Even with modern hurricane detection and warning systems in operation, the NHC can issue precise warnings only about twelve hours in advance. This should be considered against the fact that the evacuation time required for the Galveston Bay area, as an example, would be in excess of twenty hours. Down in the Florida Keys, as another example, there is only one land escape route: back up the Keys via a mostly two-lane highway with fifty bridges. It's a route that could become quickly clogged during a mass exodus. Worse, a single automobile accident could bring traffic to a standstill.

The Keys were spared a direct hit by a major hurricane, but just barely, in August 1992 when Hurricane Andrew thundered across south Florida. Andrew, a category four demon, left a wake of destruction that exceeded even that of 1989's Hugo in South Carolina. Places such as Homestead and Florida City, only twenty-five miles (40 km) north of Key Largo, were virtually leveled. (Andrew, maintaining its category four power, swirled its way into Louisiana two days later. And even before Andrew's fury was fully spent, it had become hands down the costliest storm in U.S. history.)

Storm Surges and Other Enemies

Most loss of life in a hurricane comes from the "storm surge." A storm surge is not a tidal wave, as it is often wrongly termed by the media. (A tidal wave, or tsunami, is generated by an undersea earthquake.) A storm surge is a great dome of water pushed up by a hurricane's howling winds. Tidal waters can rise twenty to twenty-five feet (6 to 7.5 m) above seal level in a matter of a few hours, with rises of several feet sometimes coming within a few minutes. At best, bridges and causeways can be inundated. At worst, as in Galveston in 1900, hundreds or even thousands of people can be drowned.

The greatest enemies of all, however, may not be the greenhouse effect and a greater number of category five hurricanes. The greatest enemies of all could be our own complacency ("it can't happen here") and skepticism (see below).

Two of the biggest media events in recent history were related to hurricanes. Television coverage of Hurricanes Gloria in 1985 and Gilbert in 1988 was almost overwhelming. Not an hour would pass without a report from a threatened area, an interview with a forecaster at the NHC, or a satellite picture of the threatening monster's cloud swirl. In truth, both Gloria and Gilbert *were* storms worthy of extensive news coverage. Gloria was a dangerous category four targeting the Eastern Seaboard. Gilbert, as previously discussed, was the most intense storm ever seen in the Western Hemisphere. Yet, in an odd and certainly unintended way, the media attention may have ultimately done more harm than good.

In the end, neither storm lived up to its ferocious billing. At landfall, Gloria weakened to a category three and crossed the coast at low tide. Damage was close to a billion dollars, but the public was left with the perception of media "overhype." It was the same with Gilbert, which not only faded to a category three after smashing Mexico's Yucatan Peninsula but didn't even hit the United States.

So, because hurricanes, especially major hurricanes, are infrequent visitors, we become complacent. And because we sometimes seem "overwarned," we become skeptical. It is a mistake, naturally. But perhaps Hugo and Andrew put things back in perspective . . . at least for awhile.

Camille

To maintain perspective, we should remember a storm called Camille. Only two hurricanes this century have crossed our coastline as full-blown category fives: the 1935 Labor Day storm and Hurricane Camille in 1969.

Camille was described by Robert Simpson, director of the NHC in 1969, as "the greatest storm of any kind ever to have affected the mainland of the United States." Camille first attained hurricane strength in the Caribbean Sea just south of the the western tip of Cuba on August 15, 1969. By August 16 the storm had moved into the southern Gulf of Mexico and was continuing to intensify. On the 17th, a U.S. Air Force reconnaissance plane ventured into Camille's eye. The crew of the aircraft measured the hurricane's surface pressure at 905 millibars (26.73 inches), which at the time was second lowest only to the Keys Labor Day storm. Camille was churning steadily toward the central Gulf Coast, and Simpson warned, "Never before has a populated area been threatened by a storm as extremely dangerous as Camille."

Camille didn't weaken before it slammed across the Mississippi Gulf Coast late on the 17th with winds estimated in excess of 175 knots (200 mph) and a swirling storm surge fifteen to twenty feet (4.5 to 6 m) above sea level. High-water marks at Pass Christian indicated water reaching over twenty-two feet (6.7 m)! Damage was almost unbelievable. The combination of screaming winds and devastating tides leveled virtually every waterfront area from southeastern Louisiana to Biloxi, Mississippi.

People caught in the storm's fury will never forget the death and destruction levied by the wind and water. And despite the warnings, death there was, for some coastal residents foolishly tried to ride out the thunderous storm. At least 144 persons perished along the gulf. Those who survived endured not only the elements but also a continuous sound level of 120 decibels, a two-hour nonstop roaring equivalent to a rocket engine test or several low-flying jets. It's a noise that embeds itself in one's memory.

Damage from Camille exceeded one billion dollars, mostly near the gulf but also in Virginia, where devastating floods resulted from the remnants of Camille several days later. At the time it was the greatest storm-generated loss in U.S. history. In every sense of the word Camille had been a truly awesome weather phenomenon. If nothing else, at least along the Mississippi shore it left a legacy of respect for hurricanes.

Coming Back to Haunt Us

Unless we've "been there"—been pinned in a building by crushing winds, been caught in the rising green-gray maelstrom of a storm surge, been witness to a roof taking flight or a house disassembling—then we tend to view such events rather academically. If indeed "what has happened, can happen," then maybe it is only natural to assume "what hasn't happened to us, can't happen." When it comes to nature, such denial does not serve us well.

Even without the greenhouse effect, even without the potential of a storm exceeding the unprecedented severity of Gilbert, the stage may be set for a tragic disaster in the 1990s. Our own complacency and skepticism could come back to haunt us. In this light, perhaps Hugo and Andrew were ghost busters, serving to lift our thresholds of complacency and skepticism. But if too much time passes before one of our coasts is again threatened by a tropical monster, the images of the wrath of Hugo and Andrew will have faded—except for those who "were there"—and we *will* be haunted.

4 / Paying the Price in the 1990s

The cost of a warmer world will be high. In the United States, agriculture will suffer the most severe consequences—at least initially, during the latter part of the 1990s. The effects of heat and drought on our food supply will burrow into the nation's economy. And as the turn of the century approaches, we could all share the burden of the greenhouse effect in the form of accelerating inflation.

Water supplies will dwindle in many areas of the country. Certainly less water will be available for irrigated farming. Stream and river flows will shrivel in some parts of the United States. Depending on where the droughts of the nineties strike, commercial river traffic could be brought to a standstill or hydroelectric generation partially curtailed. Again, many of us will share the burden: in some cities water rationing will become a way of life, and in some states recreational lakes may all but dry up.

The poor will suffer disproportionately in a greenhouse world. As summer heat grows and big cities bake, most of us will pay a little more for electricity to run our air conditioners and fans, most of us will wash our cars a little less, and most of us will shell out a few more bucks at the supermarket. But a family on welfare, subsisting in a one-room apartment in the inner city—what do they shell out? Not a few more bucks for food; they don't have a few more bucks. Wash their car a little less? What car? Run the air conditioner a bit longer? You know the answer to that. In the concrete canyons of the cities, in suffocating one-room hovels: that is where the greenhouse effect will turn deadly during the dog days of August.

Agriculture

Crop and livestock losses stemming from the heat waves and droughts of the eighties amounted to over forty-five billion dollars in the United States. There were no back-to-back droughts during the eighties, so farms and farmers had a chance to recover between sieges of dusty weather. They might be equally as lucky in the 1990s. Or they might not be.

Given the recent trend in global temperatures and considering what the computer models and analogs are telling us, there is no reason to think the next five to ten years will be any kinder to American farmers and ranchers than were the 1980s. Quite the opposite. There is every reason to think the mid- to late 1990s will be harsher.

What this might mean in terms of crop and livestock losses is anybody's guess. But let's say we're lucky. Let's say the superdrought doesn't come and we don't get consecutive years of heat and dust. Even with this "lucky" scenario, it would seem foolish to expect any fewer or any less-intense droughts than plagued us during the 1980s. Thus there is no cause to think total agricultural and livestock losses in the 1990s will be smaller than the over forty-five billion dollars in losses racked up during the eighties.

In fact, there are a considerable number of suggestions—mentioned in chapters 1 and 2—pointing at prolonged and severe drought and heat in the nineties, particularly after the cooling effects of Mount Pinatubo's volcanic haze have disappeared. Given an additional year or two of intense drought in the nineties relative to the eighties and given that farmers and ranchers, at least in certain areas of the Midwest and Great Plains, might not get a breather between virtually rainless years in the nineties, it doesn't seem particularly bold to predict crop and livestock losses exceeding sixty billion dollars during the final decade of this century.

Corn

Corn accounts for over half the total U.S. grain production. It is a crop highly dependent on July rainfall and temperature. Typically, the highest yields come in cooler-than-normal summers. In such summers corn stores more photosynthate (food), and there is usually more precipitation. During the 1930s (our analog for the latter part of the 1990s), July temperatures in the Corn Belt, from Nebraska eastward to Ohio, averaged 2° to 4° F (1° to 2° C) or more above modern means. July rainfall averaged 55 to 85 percent of current normals. The largest anomalies of heat and dryness settled over southern Minnesota, Iowa, and northern Missouri.

Since the 1930s, agricultural technology has made immense strides and crop yields have steadily increased. But even modern technology is unable to overcome the deleterious effects of severe drought. As Iowa State University economist Neil Harl points out, "We simply cannot, in all our cleverness, with our hybrids and chemicals, deal very well with adverse weather." Science is still a long way from creating truly drought-resistant plants. In 1988, drought and heat reduced midwestern corn production by 40 percent.

"You almost never gain something from nothing," says Emerson Nofziger, an agronomist with the University of Illinois Extension Service. "If you make a corn plant more drought-resistant, you usually have to give up some yield." What's worse, plants that do well in droughty years often do not thrive in "normal" years. If only farmers knew at planting time which years would be dry and which ones not. But they don't, and they won't. So they plant . . . and wait and watch.

Wheat

Wheat, the second most important U.S. grain, comprises about one-fifth of our total grain production. Seven monthly precipitation and temperature variables significantly influence wheat yields. Dryness stunts growth and hot weather encourages pests. During the 1930s, annual precipitation in Kansas, the leading wheat-producing state (winter wheat), was just 75 to 80 percent of modern means. In North Dakota, the spring wheat-growing leader, influential July temperatures averaged an astounding 4° F (2° C) higher than current normals.

Computer simulations suggest that wheat yields in Kansas, under the influence of a 1930s-type climate, would diminish an average of three or four bushels per acre over a ten-year period, about a 15 percent reduction in yield. Any single year could be significantly worse, of course. More important, researchers emphasize that computer models tend to underestimate the magnitude of weather effects on crop yields.

Soybeans

Soybeans (a leguminous grain) account for about 14 percent of U.S. grain production and 60 to 70 percent of world production. Soybeans, like corn and spring wheat, are affected by July precipitation and temperature. But unlike other grain crops, soybeans can recover from July dryness with adequate rainfall in August. Illinois is the soybean production leader. And August precipitation in and around that state averaged somewhat greater than current normals through the 1930s. Thus soybeans might turn out to be the 1990s' most resilient grain crop. On the other hand, they might not. Duke University researchers made a surprising discovery during an experiment in which they grew soybeans under conditions of increased CO_2. The soybeans grew more prolifically in such an environment, but the leaves were less nutritious; therefore insects ate more of them in order to maintain their nourishment! Which leads us nicely to the next topic: agricultural pests.

Agricultural Pests

Drought and heat affect crops not only directly—and negatively—but indirectly, too, by encouraging the growth of pests. During the 1930s the scorching heat favored wheat rust, a fungus that throughout history has probably caused greater damage than any other disease to the chief food crop. The red spores of the pest, borne by hot winds, soared over the Wheat Belt during the 1930s, and millions of tons of wheat were destroyed. Other pests took advantage of the climate change, too. The pale western cutworm thrives in warmer and drier conditions. During the Dust Bowl era the cutworm population exploded on the Canadian prairies and in Montana, decimating thousands of acres of wheat. While the cutworm proliferated in the north, the prickly pear cactus moved eastward out of Colorado. The thorny pest invaded western Kansas pastureland, and cattle production suffered greatly.

Our ability to deal with agricultural pests has improved tremendously since the 1930s. But over a decade ago a National Academy of Sciences report warned: "Our ability to solve a new problem—related strictly to climatic change and perhaps coupled with the emergence of a new pest—in time to forestall disaster is questionable." The report displayed some prescience.

In 1988 something called aflatoxin—a powerful cancer-causing toxin produced by a fungus that thrives in hot, dry weather—attacked the midwestern corn crop. Traces of aflatoxin are commonly found in many grain and peanut products, but trace amounts are not harmful. At higher levels of concentration, though, the poison becomes carcinogenic. During heat waves, corn is especially vulnerable to the toxin, since kernels crack and allow the toxin-producing fungus to flourish. Although the 1988 contamination was widespread, apparently none of the tainted corn was distributed for human consumption. (Corn is a common livestock feed, but experts say aflatoxin levels in meat eaten by humans are barely measurable.) Still, perhaps we should consider adding the aflatoxin outbreak to the warning of the 1980s.

Yes, we'll pay a price in the 1990s. Crop and livestock losses will mount, more farmers will go out of business, and food prices will jump. Yet I fear the cost will not be high enough to concern us, not high enough to frighten us. We'll take it in stride and won't feel compelled to clamor for more serious action to curtail the greenhouse effect. We will think, "After all, it's only a 'weather thing,' and we've got more important items on the agenda to worry about."

Perhaps. But then we will have lost sight of the fact that the price in the nineties is only a soft harbinger of hard times to come. Very hard times.

Water Supplies

In the drought-ridden summer of 1988 barge traffic on the Mississippi River came to a virtual standstill as the mighty waterway dwindled to its lowest level in more than a century. A river traffic jam resulted at Memphis, Tennessee, where barges were stalled, unable to deliver crops and coal.

Different drought. Different year. Different river. In 1977 the Columbia River in the Pacific Northwest, in order to sustain power generation, required emergency releases of water from British Columbia Hydro reservoirs. The great dams along the river from Grand Coulee to Bonneville were gasping for water, unable to meet their quota of electric generation without help.

In other parts of the country at other times, recreational lakes and reservoirs, for lack of rainfall, dropped to such meager levels that many boaters and fishermen abandoned them. In metropolitan areas, water restrictions went into effect. In some cities, water rationing was implemented . . . in a restaurant you had to "order" a glass of water.

Our once and future water-supply dilemma. It will all be repeated within the next ten years or so. Perhaps in different ways and in different places, but it will be repeated.

The Colorado River Basin

Let us consider in more detail the water-supply problems just one area of the country—that region of the West supplied by the Colorado River Basin—might face near the end of the 1990s. The Colorado River Basin has difficulties to begin with. With the exception of the deserts of the Great Basin, the Colorado Basin has the greatest water deficiency of any basin in the forty-eight adjacent states. (Water deficiency is defined as the average precipitation less potential evapotranspiration—soil water loss from evaporation and from transpiration of water by plants.) Yet more water is exported from the Colorado Basin than from any other river basin in the United States.

The water, tapped from rain and snow spilling off the western slopes of the Rocky Mountains, is dammed and drained at numerous points along its thousand-mile river journey from Colorado to the Gulf of California. The precious commodity is used to irrigate crops around Phoenix, keep golf courses green in Tucson, and fill swimming pools in Beverly Hills. That's right, Beverly Hills. Southern California relies almost exclusively on water imported from the Colorado River (Arizona's unused portion) and from precipitation runoff in the Sierra Nevadas of northern California. Metropolitan Los Angeles exists thanks in part to water from the Colorado River, much of which is stored in Lake Powell behind the Glen Canyon Dam in northern Arizona.

The Glen Canyon Dam marks the boundary between the Upper and Lower Colorado River basins. The Upper Basin, where most of the precipitation falls, largely encompasses eastern Utah, western Colorado, southwest Wyoming, and a small part of northwest New Mexico. The Lower Basin is predominantly Arizona and the adjacent parts of New Mexico, Nevada, and California.

Longer-range (i.e., mid-twenty-first-century) greenhouse-effect scenarios suggest the flow of the Colorado River could wither away by some 40 percent, with devastating effects on southwestern water supplies. But water supplies in the Colorado River Basin will be in trouble long before the middle of next century. They'll likely be in trouble within the next ten to fifteen years.

By law the Upper Basin must deliver seventy-five million acre-feet (maf) of water to the Lower Basin every ten years, or an average of 7.5 maf per year.* An additional 7.5 maf per ten years is required to satisfy Mexican claims. The point where the Colorado River delivery flow from the Upper to the Lower Basin is measured is at Lees Ferry, Arizona, just below the Glen Canyon Dam.

The long-term contemporary (1931–76) average yearly flow of the Colorado there is estimated at 13.5 maf. During the great drought of the

---

*An acre-foot of water is one acre of land covered one foot deep in water; it equals about 326,000 gallons, or enough for a family of five for a year.

1930s there was a significant reduction in precipitation over the Upper
Basin, and the mean annual flow dwindled to 11.8 maf, or 118 maf over
a ten-year period. Despite a 13 percent drop from the long-term average,
118 maf would provide adequate water to satisfy the potential 82.5 maf
ten-year demand in the Lower Basin. (The 82.5 maf is the 75 maf required
by law plus 7.5 maf for Mexico.)

Crisis on the Colorado

The problem is that a reduction of 13 percent is not the worst that could
happen. Dendrohydrologists have determined that during a ten-year
period in the late 1500s, the average yearly discharge of the Colorado at
Lees Ferry sputtered to 9.7 maf—a dropoff of almost 30 percent (a real 30
percent, not a theoretical 30 percent). Is it unreasonable to expect this
deficit to recur in the near future? Not at all, because once a 1930s-type
regime is established, the greenhouse effect is likely to override the
natural climatic fluctuations that bailed us out in the 1940s by returning
"normal" precipitation to the Upper Basin. The greenhouse effect, much
to the contrary, holds promise of enhanced heat and drought.

Thus, although the Colorado River Basin may survive the 1990s with
adequate water supplies, it isn't likely to make it much further than that.
Let us consider a drought scenario around the turn of the century in
which the flow at Lees Ferry dwindles to a yearly mean of 9.7 maf. Let us
also assume that hydrologists have had the foresight and luck to bring
Upper Basin water storage to its full capacity of about 30 maf before the
flow drops off. Over a ten-year period the Upper Basin must still deliver
to the Lower Basin 75 maf plus 7.5 maf (for Mexican claims), or 82.5 maf.

The Upper Basin storage (30 maf) plus the ten-year flow (97 maf)
would be more than adequate to meet the 82.5 maf requirement and still
leave the Upper Basin with 44.5 maf. *But the Upper Basin's demand for water
will exceed 44.5 maf.* By 2000, Upper Basin demand is forecast to be 5.1 maf
per year, or 51 maf over a decade; by 2010, requirements are projected to
be 5.4 maf yearly, or 54 maf over a ten-year period.

In plain language, given a severe greenhouse-effect drought—and
such a drought has precedent even in the absence of a greenhouse
climate—there wouldn't be enough water to supply Colorado River Basin
demands. It's a crisis that could plague the Southwest and its burgeoning
cities such as Phoenix and Tucson within ten or fifteen years . . . not
fifty years.

The Human Cost . . . the Dangers of Heat

Extreme heat extracts a physical toll on us in a couple of ways. First, it puts a strain on our heart. In hot weather our heart works harder, pumping more blood to the tiny capillaries in the upper layers of our skin. There the blood and thus the body are cooled by giving off excess heat to the (relatively cooler) atmosphere. When the air temperature exceeds that of our body—normal body temperature is around 98.6° F (37° C)—our body cannot shed heat through the circulatory system: there is no cooling effect if the environmental temperature is warmer than the temperature of our blood. At this point the only way we can lose heat is by sweating. The evaporation of sweat cools the skin and consequently the blood. But there are thermal limits to what our body can take, and when these limits are exceeded by very much or for very long, heat stroke and ultimately death—from heart attack or stroke—threaten. (Heat exhaustion occurs when excessive amounts of salt are lost from our body. Such a condition results from long exposure to heat or from too much activity in strong sunshine. Heat exhaustion is characterized by pale, cold, moist skin, rapid pulse and breathing, low temperature, dizziness, and a tendency to vomit.)

A second way hot weather affects us is through a change in our metabolism. Metabolism refers to the chemical processes in our body that convert food and oxygen into tissue and energy. A by-product of metabolism is heat. When the outside air temperature rises to a point where getting rid of body heat becomes a problem, our metabolic rate slows down to curtail heat production. When our metabolism becomes less active, so do other bodily functions, including our capacity to fight off infections. That may explain why summer colds are sometimes so tenacious. One study of acute appendicitis found that the disease struck twice as often during summer heat waves as during the winter.

Another danger associated with heat waves is the air pollution—smog—that often accompanies them. The stagnant air masses that frequently settle in with stifling heat allow particulates, sulfur dioxide, carbon monoxide, ozone, and so forth, to accumulate, especially over large metropolitan areas. The resultant effect on our respiratory system, at least in the long term, can be fatal.

Typically, the people who suffer most during heat waves and smog episodes are the poor. These are the persons most likely to reside in inner cities, most likely to be in poor health to begin with, and most likely to be unable to afford air conditioning.

A recent research project, however, suggests that the presence of smog and absence of air conditioning during a hot spell may be less of a factor in causing fatalities than traditionally thought. Laurence Kalkstein, on leave from the University of Delaware Center for Climatic Research and working for the Environmental Protection Agency, determined that it is a sudden jump in temperature—the temperature to which people have become acclimated—that appears to cause most problems. Kalkstein discovered, for instance, that heat begins killing people at lower temperatures in northern cities than in southern ones. Cities, depending on where they are located, have different temperatures marking the fatal threshold. For New York the tripwire is 92° F (33° C); St. Louis, 96° (36°); Dallas, 103° (39°); Los Angeles (at the international airport), 81° (27°). Cities such as Boston, Philadelphia, and Memphis are also affected; Chicago and San Francisco have a lesser problem. But many cities in the Deep South and southwestern desert areas are normally so hot that excess deaths rarely occur as the mercury soars. People in those locations are used to the heat.

Typically, Kalkstein points out, heat wave death tolls begin to rise after several consecutive days of high temperatures. By the fifth day of a hot spell, especially a hot spell accompanied by oppressive humidity, the fatality rate can skyrocket.

Air conditioning is a great equalizer, of course. Witness the large populations of elderly who thrive in hot retirement meccas such as San Antonio, Texas; Sun City, Arizona; and Palm Springs, California. They have adjusted their life-styles, as many of the rest of us will have to do through the coming decades, to hotter weather. They golf and jog in the relative cool of the early morning, run their errands and do their yard work before noon, stay indoors during the heat of the day, take a siesta, and come out again in the evening to seek a good dinner, attend a movie, or do some shopping.

New York and South Carolina

Most of us can live with hotter weather. Again, it is the poor who will suffer most with changing climate, particularly in the North, where heat waves have traditionally been less frequent. Let us consider just one example. The NASA research suggests that the number of days on which the temperature reaches or exceeds 90° F (32° C) in New York City— where the increased fatality threshold is 92° F (33° C)—will rise by about 30 percent in the latter part of the 1990s. Central Park currently averages seventeen days per year with the mercury hitting or topping 90° F. After the mid-1990s the average may well be twenty-one or twenty-two days per year. Naturally, some of the hotter summers will have a considerably greater number of such stifling days.

Remember, that will be only the beginning. NASA's work suggests that given no serious attempt to curtail the greenhouse effect, New York City by the 2030s will have a frequency of hot days (over 90° F) similar to that which Washington, D.C., now experiences. By midcentury the frequency could be more like the current norm of Charleston, South Carolina.

And if you think the United States is working diligently to prevent this kind of climatic shift from taking place, think again. (Chapter 6 will explain.)

5 / Don't Put Away Your Parka and Snow Shovel Yet

As mentioned in chapter 2, despite an overall greenhouse-effect warming trend, not all regions of the United States or the world would grow immediately warmer and drier. For instance, parts of the southwestern United States, as suggested by the 1930s analog (and the NASA model), might even experience a small drop in summertime temperatures during the 1990s (see figure 2.9). The 1930s analog also suggests that we shouldn't store away our parkas and snow shovels quite yet, either, since—rather surprisingly—there was some remarkably severe winter weather during the Dust Bowl era.

To understand why some areas of the world would grow cooler while most of the globe warms or why snowstorms and arctic outbreaks would still plague us even during years with sizzling summers, we have to know a bit about the earth's atmospheric circulation patterns. As the temperature of the atmosphere varies, so do the accompanying circulation (wind-flow) patterns. It is these patterns we must closely examine if we are to comprehend the details of how our climate will likely change under the influence of the greenhouse effect.

The Westerlies

Climate regimes are determined by the configuration of the earth's wind patterns. The wind pattern that is most important, at least for the purposes of this discussion, is one commonly called the jet stream. The jet stream is actually only a small part of the westerlies, the name given an endless band of upper atmospheric winds that circle the hemisphere at midlatitudes ($30°$ to $65°$ N).

The jet stream is the axis—actually a "tube"—of the strongest westerlies. It usually howls along at around 30,000 feet (9,145 meters) and may reach speeds up to several hundred miles per hour. In February 1982, on a TWA flight from Frankfurt, West Germany, to New York City, I gained some firsthand knowledge of how the jet stream can affect air

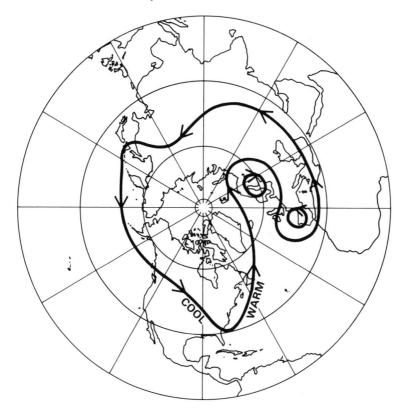

FIGURE 5.1 The undulating pattern of the Westerlies.

travel. The captain of the aircraft, a 747, said we were taking winds "on the nose of 200 knots." That meant the plane was fighting a 230-mph headwind. It also meant we would be an hour late arriving at Kennedy, even though we had taken off on time from Frankfurt. The captain, given to some hyperbole, I suspect, said, "I've been flying this route for fifty years and never seen anything like this."

In reality, the westerlies do not blow from west to east all of the time. More typically they display an undulating pattern, sweeping southward over one part of the hemisphere and back northward somewhere else. Occasionally they may form a complete loop. The undulations and loops constantly change positions, and it these shifts in the configuration of the westerlies that produce changes in our weather regimes.

Where the westerlies shift or loop northward they bring relatively warm, dry weather. Where they bend southward, cooler, wetter conditions generally result. Figure 5.1 is a schematic representation of the

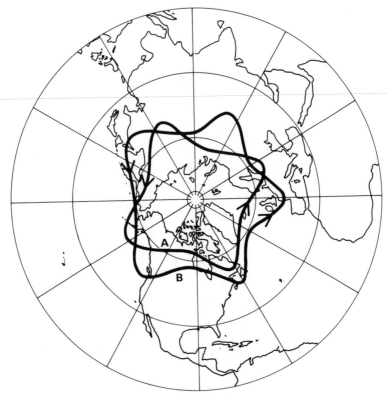

Figure 5.2 The jet stream during the Medieval Warm Period (A) and the Little Ice Age (B).

westerlies and how they influence our weather. The figure shows a looping (or blocking) pattern over western Europe. Such a configuration would produce fair weather over Scandinavia and cloudy, showery weather in Spain and France. The pattern indicated over North America would produce warm, dry weather in the western United States and cool, wet conditions in the East.

The Little Ice Age and the Medieval Warm Period

As the earth's climate warms or cools over a long period, there is a concomitant shift in the preferred pattern of the westerlies. Not only do the positions of the loops and swirls change; the actual number of such undulations may also change. Figure 5.2 gives an example. During an historical period known as the Little Ice Age, which culminated in the 1600s, the atmospheric circulation pattern was markedly different from what it was during an earlier (warm) era in the Middle Ages, from about 1000 to 1200.

The jet stream during this Medieval Warm Period (marked A in the figure) was farther north than it was during the Little Ice Age (marked B), and in the summer it displayed four basic southward bends as opposed to five during the Little Ice Age. One of the southward dips during the Little Ice Age was over the British Isles and western Europe; this implies that much cooler weather prevailed there at that time. Indeed, that was a time during which Alpine glaciers expanded to their maximum extent in historical times, burying pasturelands and passes in sheets of ice. The shores of Iceland, now ice free except for a few months each year, were locked in ice for up to six months every year.

During the warm period that preceded the Little Ice Age, the weather in Europe was extraordinarily good. Vegetation and glacier boundaries were significantly higher than even today, and grapes were grown in England and East Prussia (now largely Poland), where late spring frosts now preclude extensive vineyards. In southwest Greenland sheep and woodlands, not ice, prevailed.

Our contemporary climate is somewhere between that of the Little Ice Age and the Medieval Warm Period. Our summer circulation pattern is intermediate to the four- and five-loop systems depicted in figure 5.2. But as the greenhouse effect takes over, the pattern likely will evolve fairly rapidly to the type that dominated during the mild Middle Ages.

The reason it is important to determine which type of atmospheric circulation pattern will accompany an expected climate change is that regional differences can be more clearly defined and explained. Even in a warming world, the changes won't all be in the direction of heat and dust. For instance, during the Medieval Warm Period much of western Russia was probably relatively cool. This is suggested by the southward dip there in jet stream A, as shown in figure 5.2.

It is fair to ask—since weather records from the Middle Ages and Little Ice Age are few and far between and because we know there weren't any serfs contracted to release weather balloons into the atmosphere—how we know what type of circulation patterns prevailed then. Hubert H. Lamb, a professor with the Climatic Research Unit in England, figured it out. Acting as a climate detective, he was able to determine the probable configuration of the upper winds associated with those periods by studying the sparse records of surface temperature and air pressure available; using a deductive process involving knowledge of the various vertical relationships among temperature, pressure, and wind, he was able to reconstruct what the high-level wind patterns of those eras most likely were.

Climatic Stress

In more recent times, the westerlies reached their maximum northward extent in the early 1930s. The attendant climatic warmth of the time is reflected in the greater number of state records for high temperatures and dryness set during the 1930s than during any other decade since the 1870s, as shown in table 2.1

Table 2.1 shows something rather unexpected, too: not only were the greatest number of state records for maximum temperatures set in the 1930s; so were the greatest number of state record minima!

During the 1930s, as our climate warmed, the westerlies began to display a greater frequency of looping, or blocking, patterns. This was the start of a period of "climatic stress," as it was called by Hurd Willett, professor emeritus of meteorology at the Massachusetts Institute of Technology. When blocking patterns are established, some regions become continuously hot and dry, others cold or wet.

The high-latitude blocking configurations of the 1930s fostered persistent and anomalous weather patterns. Wind flows became "locked in" for extended periods at various times in different parts of the country. The thirties were a decade of more than record-shattering heat and drought. They were a period of remarkable cold waves and deadly floods, as well. We should keep this in mind as we move through the 1990s. Remember, the 1930s serve as the analog for the latter part of this final decade of the twentieth century.

Flakes in California, Fairbanks in North Dakota

The legacy of frigid weather during the Dust Bowl days began in January 1930, when bitter cold dominated the West. In Oregon the month brought the most persistent iciness since the late 1800s. In parts of the Willamette Valley the month was the coldest on record, and in eastern Oregon the mercury plunged to –52° F (–46° C), setting a state record that would be broken just three years later. In Montana the low reading for the month also dived to –52° F; in Wyoming, –57° F (–49° C); in Minnesota, –49° F (–45° C); in Illinois, –35° F (–37° C)—a mark that still stands; and in Oklahoma, a record –27° F (–33° C).

In 1932 a plunge of cold air into California set the stage for two unusual "snowstorms." On January 15 two inches (5 cm) of snow were judged to have fallen in downtown Los Angeles at the Civic Center. (The official measurement of snow on the ground was one inch [2.5 cm].) Snow covered the beaches at Santa Monica, and up to eighteen inches (46 cm) coated mountain passes. The only other time since 1877 that measurable snow has fallen on Los Angeles was in January 1949, when less than half an inch (1 cm) sifted down.

The other anomalous California snow event in 1932 took place in San Francisco. On December 11 almost an inch (2.5 cm) of snow fell on the downtown area. It was the most snow since 1887 and until January 1962. That snowy 1932 day was also the coldest in San Francisco history, with a high of 35° F (2° C) and a low of 27° F (–3° C). At the airport south of the city the mercury sank to an all-time record minimum of 20° F (–7° C).

In February 1933 bitter cold once again invaded the West. State records were stamped into the archives in Oregon (Seneca, –54° F [–48° C]), Wyoming (Moran, –63° F [–53° C]), and Texas (Seminole, –23° F [–31° C]). In North Dakota, the coldest blizzard known swept the state. On February 8 Bismarck had a high of –17° F (–27°C), and a low of –35° F (–37° C), and windchill factors of an astounding (and deadly) –70° to –80° F (–57° to –62° C). In Salt Lake City, Utah, the thermometer recorded its lowest reading ever (–30° F [–34° C]), and the winter itself turned out to be the coldest on record. In January, residents of Phoenix, Arizona, watched in amazement as an inch (2.5 cm) of snow blanketed the desert. That has happened only twice in Phoenix history; the other occurrence was to come in 1937.

As 1933 ended, arctic air returned, but this time to the East. A late December cold wave sent the mercury tumbling in New York (–47° F [–44° C]), Vermont (–50° F [–46° C], a state record), and New Hampshire (–44° F [–42° C]). On the night of January 28–29, 1934, another arctic front swept over the Northeast, dropping temperatures from well above freezing to well below zero (–18° C). Little did the population know that the stage had been set for the longest period of sustained cold ever to grip the region from Michigan eastward to the Atlantic coast. The unprecedented numbing weather of February 1934 was to become legendary.

On February 9, 1934, a bone-chilling –52° F (–47° C) settled over Stillwater Reservoir, New York. (It was a mark that would not be equaled until February 1979, when the thermometer again registered the same nadir, this time at a place called Old Forge). February 1934 also saw a state record established in Michigan: –51° F (–46° C) at Vanderbilt.

In major northeastern cities the bitterly cold month reduced normal commerce to a crawl. Record minima were set at Boston (–18° F [–28° C]), Providence (–17° F [–27° C]), New York (–15° F [–26° C]), Philadelphia (–11° F [–24° C]), and Buffalo (–21° F [–29° C]). The month remains the coldest ever in Boston, New York, and Buffalo. In Philadelphia it was the coldest on record until January 1977.

But the frigidity of February 1934 was soon forgotten in the heat and drought of the Dust Bowl summer that followed. The end of the year, however, brought one parting shot of icy weather, this one reaching into the orange groves of Florida, where December produced a severe freeze. Temperatures tumbled to 22° F (–6° C) in Orlando and 27° F (–3° C) in Tampa. The winter of 1934–35 was overall more moderate than its predecessor. But the winter of 1935–36 was lying in wait.

On November 30, 1935, the mercury edged below 32° F (0° C) in Langdon, North Dakota, and did not rise above freezing for the next three months . . . until February 29, 1936. For forty-one consecutive days, from January 11 to February 20, the temperature failed to struggle above 0° F (–18° C). It was the coldest extended period ever to grip any part of the United States, an ironic start for a year that was to bring the hottest summer on record to the same region.

In terms of the magnitude of the departure of temperature from normal, February 1936 was the coldest month ever in U.S. history. (On a nationwide scale, January 1979 was the coldest.) In north central Montana temperatures averaged 26° F (14° C) below normal. In fact, all of the northern Plains took on the look and feel of the Arctic tundra that bitter February. It was the coldest month known in Bismarck, North Dakota; Pierre, South Dakota; and Omaha, Nebraska. In Langdon, the mean temperature for the entire winter was –8.4° F (–22° C), the normal winter mean for Fairbanks, Alaska.

The winter was the coldest on record in Des Moines, Iowa; Kansas City, Missouri; and Minneapolis, Minnesota, as well as in Pierre and Omaha. Minneapolis suffered through thirty-six straight days with subzero (–18° C) minima, and the temperature for February averaged a flat zero. Only January 1977, with a mean slightly above zero, has since come close to the tenacious iciness of February 1936.

The winter saw state marks for extreme cold established in North Dakota (Parshall, –60° F [–51° C]) and South Dakota (McIntosh, –58° F [–50° C]), with city records shattered in Bismarck (–45° F [–43° C]); Lander, Wyoming (–40° F [–40° C]); and Denver, Colorado (–30° F [–34° C]).

In the autumn of 1936 Denver was buried by a record early snowfall of over twenty-one inches (53 cm) in late September. It was the harbinger of another wicked winter in the West.

And wicked it was. January 1937 established records for monthly cold all over the West, from Montana to Arizona, from California to Colorado. In Phoenix, for the second time within four years, an inch (2.5 cm) of snow fell, an event that has not been repeated since. In California, severe freezes plagued the citrus crops, but the trees survived. Boca, in the northern California Sierras, tallied a state record of –45° F (–43° C), and San Jacinto, Nevada, set a state mark of –50° F (–46° C).

While cold and snow blitzed the West, rain and warmth invaded the East. Record floods surged through the Ohio and mid-Mississippi rivers, and unusual warmth wiped out the winter sports season in New England. In Vermont, only a little over a foot (30 cm) of snow fell in January and February, and monthly temperatures in January ran almost 9° F (5° C) above normal. In Boston, Massachusetts, residents needed brooms, not shovels, to rid the area of the paltry nine inches (23 cm) of snow that wafted down that winter. Normal Boston snowfall is about forty-four inches (112 cm) per winter.

In late 1937, wintry weather ignored the West and pushed into the Southeast. In December, what was termed a "moderate" freeze frosted the oranges in Florida. Orlando dipped to 25° F (–4° C) and Tampa to 28° F (–2° C).

The anomalous 1930s cold spells closed out in 1938 with a parting shot at the Northeast. An early winter arctic outbreak pushed temperatures below zero (–18° C) in late November—the earliest ever in Albany, New York, and Burlington, Vermont.

Thus despite an overall warming trend, the Dust Bowl days brought periods of numbing cold. Icy winters were particularly persistent in the West, notwithstanding a greater frequency of hot summers. This is suggested in figure 5.3: The figure indicates that not all parts of the United States were warmer during the 1930s (relative to current means). The reasons for this are discussed in more detail in chapters 9, 10, and 11. Suffice it to say here that it has to do with the way the configuration of the westerlies changes with a warming climate. The shifting westerlies also bring altered precipitation patterns. So, although much of the United States was relatively drier in the thirties, a few places were actually wetter. Look again at figure 2.10.

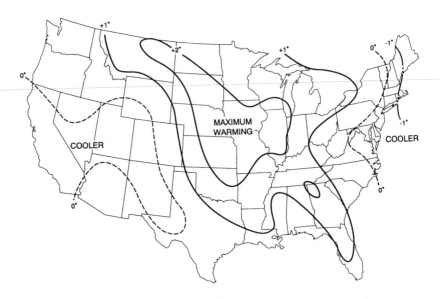

FIGURE 5.3 The mean annual temperature differences between the 1930s and now.

Drownings in the Dust Bowl

While the Midwest dried up under the onslaught of great droughts, record rains and floods decimated other regions. But even the Great Plains were subject to occasional intrusions of significant amounts of moisture. Before things really fell apart, agriculturally speaking, copious August rains in 1932 produced a record corn crop in Iowa. In 1935, excessive precipitation in the Republican and Kansas River basins led to extreme flooding in Colorado, Kansas, and Nebraska. Railroads were washed out, and 110 people drowned, a true irony of the Dust Bowl era.

Floods in southern California in the 1930s took at least 146 lives. The worst of the washouts came in February 1938 when a series of storms dumped up to thirty inches (762 mm) of rain on the San Bernardino and San Gabriel mountains. Huge floods roiled down the San Gabriel, Santa Ana, and Mohave rivers. The death toll reached eighty-seven, and the property damage tally was over seventy-eight million dollars.

Just to the south of the heart of the Dust Bowl, inadequate precipitation was not a problem during the 1930s. Texas generally had greater than normal annual precipitation and occasionally had to battle the results of too much rainfall. Floods came along fairly regularly in the mid-

and late 1930s. One of the worst covered the state in September 1936. Torrential downpours kicked off widespread flooding, and a twenty-five-inch (635-cm) cloudburst at San Angelo washed away 300 homes.

Great precipitation contrasts occurred in the Pacific Northwest during the 1930s. In Washington state, the wettest year on record followed by just one year the driest ever. At Wahluke, in eastern Washington, a total of 2.61 inches (66 mm) of precipitation in 1930 was just barely enough to kick up dust. But in 1931, 184.56 inches (4,688 mm) cascaded down on the Wynoochee and Oxbow areas of the Olympic rain forest.

In Oregon, the greatest and least annual precipitation totals were separated by only two years. In 1937, 168.88 inches (4,290 mm) of rain and snow (melted equivalent) pounded down on Valsetz in the Coast Range. In 1939, the Warmsprings Reservoir, in the lee of the Cascades, was sprinkled by only 3.33 inches (85 mm).

In the eastern third of the nation, the story was not so much diversity in precipitation amounts as it was too much precipitation.

In March 1936 unprecedented floods thundered through New England. Alternate periods of rain and thaw early in the month softened a vast blanket of ice and snow across the region. Then, on March 12, heavy rains pelted down on the soggy mixture. Rapid snowmelt and thawing ice added to the saturation. An immense volume of water sluiced into streams and rivers, and chunks of ice soon clogged swollen waterways. Rivers burst over their banks, smashing bridges and dams and submerging towns. Most means of communication—highways, railroads, and wires—were severed.

As the mighty surge of water swept toward the sea, the Merrimack Valley of New Hampshire and Massachusetts was especially hard hit. Water eighteen to twenty feet (6 m) deep—13.5 feet (4 m) higher than any previous flood crest—churned through the main street of Hooksett, New Hampshire. In the Massachusetts cities of Lawrence, Haverhill, and Lowell, mills and factories were totally inundated. In Maine, eighty-one bridges were swept away as the three principal drainage basins, the Penobscot, the Kennebec, and the Androscoggin, all reached record levels.

Virtually the entire length of the Connecticut River reported unprecedented flood heights. At Hartford, where flood records had been maintained for more than 300 years, the 1936 crest topped the list by 8.6 feet (2.6 m) at 37.6 feet (11.5 m). Much of the city was under water. Later crests, from the Great New England Hurricane in 1938 and Hurricane Diane in 1954, would approach but never exceed the 1936 mark.

Floods also swirled through Pennsylvania in March 1936, forcing the

Susquehanna River at Harrisburg to 3.5 feet (1.1 m) over its previous crest. (In 1972, Hurricane Agnes generated floods that topped the 1936 crest by four to six feet [1.5 m].)

In all, 107 lives were lost in the Northeast, and property damage reached 270 million dollars. (By contrast, Agnes left over three billion dollars' worth of destruction.)

Rampaging floods returned in 1937, but this time to the Ohio and Mississippi rivers. A month of heavy rains throughout the Ohio Valley reached a climax in late January, the same month the Far West was battling record cold and snow. From January 22 to January 26 the Ohio River was above flood stage along its whole 1,000-mile (1,600-km) length from Pittsburgh, Pennsylvania, to Cairo, Illinois. At Louisville, Kentucky, all previous flood stages were exceeded by ten to eleven feet (3.2 m); about two-thirds of the city's residential areas and almost all of its business district were awash. At Cincinnati, Ohio, the water surged almost nine feet (2.7 m) higher than any previous crest, and 10 percent of the city was submerged.

In southern Illinois, about 90 percent of Gallatin County was covered by water. Nearly all of the 35,000 residents of Paducah, Kentucky, were forced to evacuate. Some river towns were completely abandoned, later to be rebuilt on higher ground. South of Cairo, on the mid-Mississippi River, new high-water marks were set for 350 miles (560 km).

In all, the floods along the Ohio and Mississippi rivers that January claimed 250 lives and left property damage totaling 470 million dollars. At the peak of the disaster, 12,700 square miles (32,900 square km) of land and 75,000 homes were under water.

The irony of the hot-and-dry Dust Bowl: floods and freezes. But the implication is that the journey to a warmer world is uneven and contradictory. Heat waves and droughts will, in all probability, capture the headlines in the nineties, but frigid spells and floods will share the media spotlight. And not all parts of the country will experience a trend toward warmer and drier weather as we approach the end of the century; a few places will turn a bit cooler and a little wetter. Not even all years will participate in the march toward a greenhouse world. Keep in mind the predicted global cooling through 1993, courtesy of Mount Pinatubo's volcanic haze. The danger is that after any such cooling we might be tempted to think, "Well, the scientists were wrong, the greenhouse effect isn't real . . . isn't important."

We would be wrong, of course. The wolf would only be napping.

PART II

Greenhouse Gases, Climate Models, and Global Forecasts

6/ Pogo Was Right ("We've Met the Enemy, and He Is Us")

Carbon dioxide is not in itself harmful. It is found naturally in the atmosphere. By volume it comprises about 0.03 percent of the air we breathe. The two largest constituents of the atmosphere are nitrogen and oxygen. Nitrogen accounts for about 78 percent of the atmosphere's composition, oxygen for just a little under 21 percent.

Carbon dioxide is what we exhale when we breathe. Plants and trees use CO_2 in the photosynthesis process, the process that combines CO_2 and water in the presence of chlorophyll and sunlight in order to manufacture carbohydrates. All life depends upon photosynthesis, either directly or indirectly. That is because all animals feed either on plants (carbohydrates) or on animals that eat plants. Obviously we need CO_2.

We need it for another reason, too. In terms of temperature, it makes our planet habitable. Carbon dioxide has physical properties such that it is relatively transparent to solar radiation—sunshine—but relatively opaque to the earth's heat radiation. Or, saying it another way, CO_2 allows sunshine to heat the earth but then traps much of the heat near the earth's surface, rather than permitting it to radiate back to space. This greenhouse effect warms the earth. (Actually, the term *greenhouse effect*, though in common usage, is a bit of a misnomer. A real greenhouse is effective for reasons in addition to its green glass acting like CO_2; a real greenhouse physically prevents outside air from entering the greenhouse.)

Too Much of a Good Thing

The earth's greenhouse effect helps maintain the global mean temperature at a comfortable 59° F (15° C) . . . the average temperature of Raleigh, North Carolina. If we did not have CO_2 and various other greenhouse trace gases, our planet would be a sparsely populated ice ball with an average temperature of 0° F (–18° C). In Canada's Northwest Territories the village of Alert, north of the Arctic Circle, has a mean temperature of 0° F. Wintertime readings there dip to –50° F (–46° C) or –60° F (–50° C), and even in midsummer, subfreezing nights are common.

Of course, there can be too much of a good thing. The planet Venus, for instance, has an atmosphere choking in CO_2. As a result the average temperature of Venus is about 750° F (400° C), fine for broiling meat but not for raising families. (Without CO_2 Venus would be a frozen planet with a mean temperature near –60° F [–50° C]).

But being a frozen planet is not in the earth's immediate future. Quite the opposite. We're heading in the too-much-of-a-good-thing direction. Owing primarily to our ever-increasing combustion of fossil fuels, we're injecting more and more CO_2 into the air every year. Fossil fuels are products of the fossilized remains of plants and trees. When we burn coal and oil, the CO_2 that was absorbed by plant life eons ago is released back into the atmosphere. But it is released at a much faster rate than plant life can use it. The result is that the concentration of atmospheric CO_2 annually grows by several tenths of 1 percent. Altogether, the air now holds over 27 percent more CO_2 than it did just a little over 100 years ago.

We continue burning greater amounts of fossil fuels each year—and releasing greater amounts of CO_2 into the atmosphere—because the earth's burgeoning population demands more and more energy. Our appetite for fossil fuels has doubled since 1960 and grown tenfold since 1900. Driven by our quest for energy, we currently spew six billion metric tons of carbon into the atmosphere every year.

Spying on Carbon Dioxide

Scientists have been carrying out actual measurements of airborne carbon dioxide for only a short time, since the late 1950s. Continuous measurements were begun in 1958 at an observatory on the dormant Hawaiian volcano, Mauna Loa. The concentration of CO_2 that year was pegged at 315 or 316 parts per million (or ppm, a chemical measure). Since then the National Oceanic and Atmospheric Administration has expanded its CO_2 monitoring program, and there are now two dozen sites around the globe that keep track of atmospheric CO_2. The Mauna Loa record has been essentially duplicated by other measurements taken in such places as American Samoa; Barrow, Alaska; and Antarctica. By May 1990 the global concentration of CO_2 had grown to 357 ppm. Figure 6.1 shows the increasing trend of airborne CO_2 concentrations from 1958 to 1990.

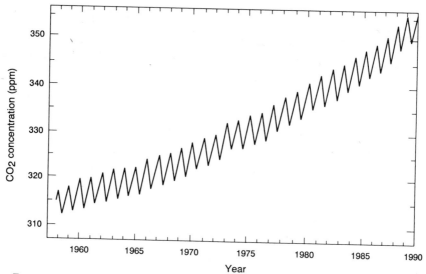

FIGURE 6.1 The trend of atmospheric carbon dioxide concentration since 1958.

The figure also shows an annual up and down variation in CO_2 of about 5 or 6 ppm. This is superimposed on the overall, long-term upward trend, and reflects a seasonal oscillation. During summer, when plants and trees are actively growing and using carbon dioxide in the photosynthesis process, a great deal of CO_2 is being consumed. In winter, when many plants are dormant, less CO_2 is being used, and more of it remains in the atmosphere. In the Northern Hemisphere, which has most of the world's flora, atmospheric CO_2 reaches a peak just after winter, usually in April or May, and falls to a minimum at the end of summer, usually in late September or October.

Although CO_2 monitoring has been going on for only about three decades, scientists have a good idea of how much carbon dioxide was in the air before they started measurements. By knowing the amount of fossil fuel burned annually, researchers have been able to reconstruct CO_2 concentrations back to the beginning of the Industrial Revolution, when the use of fossil fuels, coal in particular, began to accelerate. This occurred around 1860, and scientists estimate that at that point the atmosphere contained roughly 280 ppm of CO_2. Further research—analyses of air trapped in ice cores—suggests that CO_2 concentrations underwent little variation for centuries prior to that. By the end of this century, atmospheric CO_2 amounts will be near 375 ppm and still climbing.

Carbon Dioxide and Rain Forests

As just discussed, plant life plays an important role in the CO_2 cycle. This is one reason why the destruction of the world's tropical rain forests is so frightening. By destroying these forests we are not only releasing an additional one to three billion metric tons of carbon into the air each year, we are forcing CO_2 to remain in the atmosphere rather than permitting it to be consumed by the rain forests as part of the photosynthesis process. In effect, we are turning the earth's tropical forests into a net source of CO_2 when they should be a storehouse.

Tropical woodlands, though they cover only 8 percent of the earth, contain almost half the live wood on our planet. As a carbon storehouse they extract carbon dioxide from the atmosphere (through photosynthesis), emit oxygen, and store the carbon in their wood, leaves, roots, and surrounding soil. But these great forests are currently being obliterated at an astounding rate. Nowhere is the problem more acute than in Brazil's Amazon Basin.

The Amazon rain forest is immense, covering an area 90 percent the size of the contiguous United States. Yet by one estimate, in 1987 alone 120,000 square miles (310, 880 square km) of the forest were wiped out by burning, logging, mining, and flooding (the result of hydroelectric projects). This destruction may have contributed as much as 10 percent of the increase in atmospheric CO_2 that year. Most of the devastation is caused by fire, as peasants stream from the squalor of overcrowded cities and torch vast reaches of forest to create subsistence farms and ranches. You can hardly fault them. As Henry R. Breck, a trustee of the Natural Resources Defense Council, points out, "Whether it is in Brazilian, Philippine, African or Indonesian forests, we must concede that the people actually carrying out the destruction are obeying the first rule of any living being: to survive."

The irony of this is that the period of survival probably is limited. Rain-forest acreage burned for cropland is of limited use. The soil is thin and poor, and after about fifteen years crops will no longer grow. The forest will not regenerate, either. And if a high rate of destruction continues there will be no Amazon forest left shortly after the turn of the century. We will have managed to wipe out fifty million years of evolution—including plant and animal species—in a little over fifty years. It brings to mind a prediction by Albert Schweitzer: "Man has lost the capacity to foresee and to forestall. He will end by destroying the earth."

Happily, through the concerted efforts of conservationists and other activists (see chapter 13), the rate of Brazilian forest destruction did decline significantly after 1987. But on a global scale, the news is not so good. Recently the United Nations Food and Agricultural Organization reported that tropical rain forests worldwide are being leveled 50 percent more rapidly than a decade ago. A forested area the size of Washington state vanishes every year.

Other Culprits

On a global scale, carbon dioxide is still the single biggest contributor to the greenhouse problem. In the United States, three-quarters of the CO_2 shoved into the atmosphere each year comes from electric utilities, gasoline-powered transportation, and homes. Coal- and oil-fired utilities are the biggest single source of fossil-fuel emissions, accounting for one-third of the country's CO_2 contributions. And although CO_2 remains the prime "heavy" in the greenhouse-effect drama, other culprits have recently been identified.

Cattle and Termites

Cattle and termites? Well, only indirectly. The real culprit here is methane, a gas produced—among other places—in the digestive tracts of cattle and termites. Termites, says Patrick Zimmerman of the National Center for Atmospheric Research, "could be responsible for as much as 50 percent of the total atmospheric methane budget." (Most other researchers peg the figure at 5 or 10 percent.) Termites manufacture enormous amounts of gas as they feed on dead vegetation, and dead vegetation has been proliferating in . . . you guessed it . . . tropical rain forests. The actual burning of such forests also releases methane.

Livestock, cattle in particular, contribute to methane "pollution" primarily through manure and flatulence. I'm not sure whether that is being measured or not, but the total contribution of methane to the greenhouse effect is. Whereas CO_2 accounts for a little over half the problem, methane is responsible for about one-eighth.

Another important source of methane is the decay of organic matter in rice paddies, natural wetlands, and peat bogs. Altogether, atmospheric methane has been increasing 1 to 2 percent annually. It is clear that the processes which produce methane are largely natural and that they have long been part of the climatic equilibrium. We obviously aren't going to stop growing rice or raising cattle, but we *could* markedly slow the destruction of tropical forests. This would not only provide less fodder for termites but would also directly slow the release of CO_2 into the air.

Natural processes, however, may not be the only way methane gets into the atmosphere. Recent studies suggest that man himself, through the use of nitrogen-based fertilizers and the production of acid rain, may be contributing unwittingly to the increase of atmospheric methane. Paul Steudler and several of his colleagues from the Woods Hole Marine Biological Laboratory in Massachusetts carried out some experiments with soil and nitrogen. They discovered that certain kinds of soil bacteria exposed to the nitrogen in fertilizers and acid rain consume the nitrogen—rather than methane from the air as they would normally—to generate energy needed for growth. In other words, the soil bacteria show a preference for nitrogen introduced into the soil, over naturally occurring atmospheric methane. The end result is that less methane is taken from the air, and the atmospheric concentration of methane increases more rapidly than it would in the absence of nitrogen fertilizers and acid rain.

Fertilizers

Nitrous oxide, like methane, is another potent greenhouse gas whose effect is currently small—around 6 percent—but growing. About one-third of the nitrous oxide added to the air each year results from human activity, largely the use of nitrogen-based fertilizers. And about half that contribution comes from the United States, China, and Russia. The combustion of fossil fuels, including gasoline, is an additional source of nitrous oxide.

Ozone

Ozone, as you probably recognize from reading about holes in the ozone layer fifteen miles (25 km) above us, is not an environmental "culprit," but the lack of it is. With diminished ozone overhead, more ultraviolet radiation from the sun reaches us and threatens an increased risk of skin cancer and other health problems.

There is a "but" clause to this, however . . . but, in the *lower* atmosphere, ozone is a greenhouse-effect culprit! It is speculative at this time, but scientists think that ozone—produced when photochemical smog (as in Los Angeles) is formed—and other miscellaneous trace gases, such as upper atmospheric water vapor, could be important contributors to the greenhouse effect. Ironically, researchers also think the loss of ozone in the *upper* atmosphere may have a cooling effect on the earth; that is, it may be acting as a greenhouse-effect mitigator.

Air Conditioners

Air conditioners and refrigerators use something called chlorofluorocarbons (CFCs) as coolants. One of the most popular brand names is Du Pont's Freon. In 1975 a University of Chicago scientist, Veerabhadran Ramanathan, was amazed to discover that Freon was a surprisingly effective greenhouse gas. Today, CFCs account for one-quarter of our greenhouse gases, and they've been growing at around 4 percent annually.

CFCs have uses other than as refrigerants. They help make the plastic foam used in home insulation and disposable food containers . . . hot drink cups and hamburger holders. CFCs are also important as spray can propellants, although the United States, Canada, Norway, and Sweden banned them for that use in the late 1970s. In a more recent application, CFCs are employed as solvents to clean computer chips.

When they were developed in the 1930s, CFCs seemed like a chemist's dream: odorless, nonflammable, noncorrosive, and nontoxic. Chemists weren't aware then that in the upper atmosphere CFCs undergo a werewolflike transformation. Triggered by ultraviolet light from the sun, this transformation releases chlorine atoms from the CFCs. The chlorine then goes about merrily gobbling up ozone at a prodigious rate; for each chlorine atom freed, 100,000 molecules of ozone are eaten. The result: holes in the ozone layer.

That CFCs were the cause of our tattered ozone layer was finally confirmed by scientists in 1985. In 1987, thirty-seven nations signed the Montreal Protocol, which called on most signatory countries to reduce CFC production and consumption by 50 percent before the year 2000. In 1990, as news about the ozone layer became grimmer, the protocol was amended to require a complete phaseout of CFCs by industrial nations by 2000 and by developing countries by 2010. (A fund of at least 160 million dollars was to be established to aid developing nations in finding substitute technologies.) Ninety-three countries agreed to the amended protocol. (The United States, in 1992, spurred by ever-more-alarming reports on ozone depletion, took unilateral action to mandate phaseout of CFCs by the end of 1995.)

At least the Montreal Protocol shows that nations can work together for a common environmental cause. Keep in mind, though, that this cause is a depletion of the ozone layer (with globally uniform consequences), not the greenhouse effect (with globally diverse consequences). Keep in mind, too, that despite the apparent good will of companies like Du Pont, which plans to phase out the manufacture of CFCs shortly, there is no corporate benevolence involved. The *Miami Herald* (October 16, 1988) reports that Du Pont will realize an additional 1.8 billion dollars in profit from its sale of CFCs prior to the time a substitute is available. In essence, consumers will foot the bill for Du Pont's research and development of a CFC replacement. Other firms have already developed solvents to replace CFCs used to clean computer parts and alternatives for plastic-foam fast-food containers. You can bet these companies will not lose money.

Unfortunately, CFC phaseout may have a negligible effect on curtailing greenhouse warming. For one thing, although CFCs are indeed greenhouse gases, their elimination means the ozone layer will begin repairing itself; that in turn means the cooling effect of ozone loss in the upper atmosphere will diminish. Scientists currently feel that CFC-induced warming (in the lower atmosphere) just about equals CFC-induced cooling (because of ozone loss in the upper atmosphere). For another thing, one of the early CFC replacements, while not an ozone destroyer, remains a contributor to the greenhouse effect! Still, *something* to address an environmental threat was done. And that's more than has been done in the United States to deal with the greenhouse effect.

Researchers hope, however, that the ozone crisis will serve as a wake-up call for us vis-à-vis the greenhouse effect. As Michael Oppenheimer, senior scientist with the Environmental Defense Fund, points out, "The effects of CFCs were not manifest for years, but when they came they fell like a ton of bricks. The same could be true with global warming. With both, there's a nasty combination of lag and irreversibility." (Even if the release of CFCs ceased immediately, the ozone layer would not repair itself before the middle of next century.)

U.S. Inactions in the 1980s

In *The Greenhouse Effect* (1980) I suggested that the United States should establish certain energy goals for the year 2000. Attainment of these goals, I reasoned, would not only address the greenhouse effect but would also significantly reduce our reliance on imported oil. In 1977, 47 percent of our petroleum came from foreign suppliers. Thus any actions to reduce imports made sense even in the absence of a greenhouse threat. A greenhouse world is a global problem, of course, and the United States alone can't solve it. But we could lead an attack on the dilemma. Sadly, though, for over a decade the United States has remained mired in political inaction relative to the greenhouse effect and energy policies. It is clear even now that the goals I hoped for by the end of the century will not come close to being reached.

Energy Conservation

In one area, however—energy conservation—the United States has done fairly well, though not nearly so well as many had hoped. In my energy goals for 2000 I targeted an admittedly difficult 20 percent reduction in consumption. But it's obvious we aren't ready for the kind of sacrifices such a large drop in energy use would require. And what gains in conservation we have made have not been inspired by the greenhouse threat.

Between 1973 and 1985, spurred largely by huge oil price hikes imposed by the Organization of Petroleum Exporting Countries (OPEC) in 1973 and 1979 and by the recession of the early 1980s, U.S. energy consumption actually declined slightly. At the same time our gross national product rose 30 percent in real terms, indicating we were using energy more efficiently. Then came the economic recovery and the petroleum glut of the late 1980s. U.S. energy consumption rates responded by rising to pre-OPEC-price-shock levels, around 3 per cent per year. Oil imports rose, too: from a minimum of 27 percent in 1985 to 42 percent in 1990. Predictions are that petroleum imports will reach the 50 to 60 percent range by the end of the century. (The ramifications of this are discussed in chapter 12.)

Solar Energy

Along with a 20 percent reduction of energy consumption, I recommended in *The Greenhouse Effect* that we shoot for a 25 percent contribution from "clean" solar energy by 2000. Solar is defined broadly here to include wind, hydro, ocean, and biomass. Even President Jimmy Carter, in mid-1979, set a national goal of 20 percent for solar energy by the end of the century. We will fail miserably. President Ronald Reagan slashed the budget for the development of solar energy by 80 percent, state and federal subsidies for its use have been cut, and electric utilities—which have billions of dollars invested in other sources—aren't about to become solar trailblazers. The amount of U.S. commercial energy supplied by the sun has remained virtually constant since 1977 at about 6 or 7 percent.

Nuclear Power and the Law of Unintended Consequences

In *The Greenhouse Effect* I counseled that with a significant reduction in overall energy consumption, nuclear power should be able to supply about 16 percent of all our energy needs by 2000. Before the accident at Pennsylvania's Three Mile Island nuclear power plant, President Richard Nixon's energy planners had foreseen "nukes" supplying 40 percent of all U.S. electricity by 2000. Subsequently President Carter's administration projected that no more than about 25 percent of our electricity, or a bit less than 8 percent of our total energy, would come from nukes by the end of the century. We won't even make that goal.

Nuclear power currently supplies 20 percent of our electricity, or 6 to 7 percent of all our energy. Due to the strident and largely emotional opposition to nukes, no new nuclear energy plants have been ordered since 1978. Between 1980 and 1984, fifty-three plants were canceled at thirty-one sites.

In his book *In Search of History* Theodore White writes of what historians call the Law of Unintended Consequences. "The Law of Unintended Consequences," he says, "is what twists the simple chronology of history into drama. The operation of this law, as instrumented by Americans in the early postwar years . . . is classic. Both in Asia and Europe, we were bound . . . by gratitude to old allies—England and China. On both continents our enemies, the Germans and Japanese, were to be treated with the utmost severity. . . ." Under the Marshall Plan, the plan to rebuild Europe after World War II, the American attitude toward the Germans was "make these bastards work their way back."

In Germany, White watched as the U.S. Army forced the Germans to work a forty-eight-hour week and to rebuild their factories, roads, and bridges; all this before new housing and clothing were provided. Yes, we made them earn their way back. And the Japanese, too. No one could have envisioned that forty years later Germany and Japan "would threaten, like giant pincer claws, America's industrial supremacy." The Law of Unintended Consequences had come home to roost.

And so it seems to be with nuclear power. Certainly environmentalists and other opponents of nuclear energy are well-intentioned. Their intentions, if not founded more on fact than emotion, are at least focused on saving us from a hypothetical nuclear disaster. But in their zeal to halt the expansion of nuclear power in the United States, opponents have, at the same time, permitted the level of fossil-fuel use to remain virtually unchanged. In 1977, 91 percent of our energy was supplied by fossil fuels. Fifteen years later the figure had dropped only slightly, to 85 percent. And embedded in that figure is an expanded role for coal relative to oil, and coal is the dirtiest of all the fossils in terms of CO_2 pollution. (In *The Greenhouse Effect* I had set a goal of 59 percent for fossils by 2000, with the use of coal being cut by 35 percent.)

The Law of Unintended Consequences. In their efforts to save us from one environmental threat—the largely imagined ravages of nuclear power—nuclear opponents have left the door open to an even greater environmental threat: the greenhouse effect.

And it's for real. Not imagined. Not hypothetical.

Taking the Earth's Temperature

So Pogo, the beloved cartoon possum with a penchant for political allegory, was right: "We've met the enemy, and he is us." Pogo wasn't thinking about the greenhouse effect, but he could have been. (In fact, Pogo *has* begun thinking about the greenhouse effect. One day in early 1989 as he drifted over the Okefenokee in his swamp boat, he mused, "mebbe that *poet* was *right* . . . mebbe this *is* the way the world ends . . . not with a bang but a wither." Apologies to T.S. Eliot, I'm sure.)

Despite over a decade of opportunity to do something positive about our energy future, if not the greenhouse effect, we've stood by and watched our oil imports soar and the earth's temperature continue to climb.

Consistent with the amount of greenhouse gases, primarily carbon dioxide, that we have put into the atmosphere since the beginning of the Industrial Revolution, we have managed to warm our planet about 1° F (0.6° C) over the past century. To be honest, scientists aren't sure if all this warming can be attributed to the greenhouse effect. But in the 1980s, as greenhouse gases continued to flood into the air and global temperatures reached record levels, researchers such as James Hansen stated flatly that "we can ascribe with a high degree of confidence a cause and effect relationship to the greenhouse effect."

Hansen and other NASA researchers study temperature records that date back to 1880. The records are from continental and island locations (currently about 650 sites) that in recent decades have represented 70 to 80 percent of the globe. Around 1900 about 50 percent of the earth was represented.

Other scientists, using different approaches—such as examining data from upper air observational equipment (radiosondes)—and different sources—such as marine data (temperatures from ships)—have essentially replicated the temperature trends found by NASA. British researchers employ shipboard-observed ocean temperatures as well as land-based readings in their analyses.

The warming of the earth definitely is real.

7 / How Do We Know What's Going to Happen?

Weather and climate are not entities that can be studied handily in a laboratory. The only laboratory available is the earth's atmosphere, and the only way we can study in detail what's going on is in "real time," that is, by analyzing events as they happen. We can't carry out large-scale experiments in weather and climate and then return the atmosphere to its original state. Once we change something, it's changed. As it turns out, we're inadvertently performing an experiment of unprecedented proportions—and consequences—with atmospheric CO_2. We're calling the experiment the greenhouse effect.

Scientists don't know *exactly* how this experiment is going to turn out, but they have a pretty good idea. Remember, Svante Arrhenius did some calculations back in the late 1890s (see chapter 1) and predicted that a doubling of atmospheric CO_2 would lead to global warming on the order of 9° F (5° C). When Syukuro Manabe and Richard Wetherald ran their computer model in 1967, they came up with an answer of 4° F (2° C) for doubled CO_2.

When *The Greenhouse Effect* came out in 1980, the consensus among researchers was that doubling the amount of CO_2 in the air would lead to warming in the range of 3.6° to 5.4° F (2° to 3° C). Since then computer models have become more sophisticated, but the answers they are giving haven't changed much. The current generally accepted range for greenhouse warming resulting from doubled CO_2 is 3° to 8° F (1.5° to 4.5° C). Curiously, these figures come close to bracketing the original numbers of Arrhenius, Manabe, and Wetherald. Now you can begin to understand what Stephen Schneider means when he says, "The greenhouse effect is the least controversial theory in atmospheric science."

Perhaps what you can't understand is how scientists know how much warming might take place in response to increasing CO_2 and various trace gases. Unlike the increase in CO_2 itself (and the increase in trace gases), greenhouse-induced warming is not something we've been able to measure directly. This is because, up until now, the warming has not been large enough to distinguish from the natural fluctuation of global temperature.

Climate Models

Thus without a clear-cut greenhouse-effect "signal" (temperature trend) to track and lacking the benefit of a "laboratory" in which to carry out experiments designed to predict the future, researchers have been forced to produce representations of atmospheric behavior using mock-ups, or models. These models are constructed with numbers and complex mathematical relationships. Then, with the aid of very large and very fast computers, the models can be used to examine atmospheric response to various influences, such as increasing amounts of CO_2 and trace gases.

Underlying all climate models are the fundamental laws of physics that govern the behavior of the atmosphere. If you took high school or college physics, you may recall that these laws include Newton's laws of motion; Charles's and Boyle's laws relating pressure, temperature, and density; the laws of thermodynamics (relating heat to other forms of energy); and the principle of continuity (which says mass is neither created nor destroyed). Actually, you don't have to remember these laws at all (I don't), but it is important you know they exist and that climate models are based on them.

In truth, climate models are relatively primitive. Scientists cannot possibly hope to represent all of the intricacies and nuances of real atmospheric processes. Even the most complex climate models have serious shortcomings. One of the greatest weaknesses has been that types and amounts of clouds generally have not been computed. (Until the advent of the NASA model—which will be discussed later in this chapter—clouds were accounted for in an "average" way by adjusting incoming and outgoing radiation.)

"Clouds are an important factor about which little is known," says Schneider. "When I first started looking at this in 1972, we didn't know much about the feedback from clouds. We don't know any more now than we did then." Feedback refers to the propensity of an entity to exacerbate or mitigate a particular trend. For instance, feedback from clouds might either intensify or partially negate greenhouse warming (but would not reverse it, since the feedback is driven by the warming itself).

Another major deficiency in climate models is that interactions between the atmosphere and the oceans have not been adequately represented. Such interactions are important, because the oceans store and transport large quantities of heat and thus exert significant long-term effects on weather and climate.

Schneider thinks the next major step in improving climate models will be to take a better look at what the oceans might do in response to warming. "You need to couple a three-dimensional atmospheric model with a three-dimensional ocean model; a model with winds at the top and ocean currents and salinity at the bottom," he says.

Researchers, by knowing how much fossil fuel is consumed annually, have a very good idea of how much CO_2 is being released into the atmosphere by people every year. However, since 1958 only about half of the CO_2 known to have been produced by fossil fuel consumption has shown up in the atmosphere. Most of the remainder has been absorbed by the oceans, much of it in deep water layers. The oceans are thus a tremendous "storage locker" of CO_2.

The oceans also, since they warm (and cool) more slowly than land masses, effectively delay the full magnitude of global warming. This is why the modeling of ocean-atmosphere interaction is so important. (Current climate models account for the warming delay only crudely.) The fear is that something may happen to trigger a positive feedback— one that would greatly enhance the greenhouse effect. Wallace Broecker, a geochemist at Columbia University's Lamont-Doherty Geological Observatory, argues that based on experiments with newer ocean-atmosphere models, there is clear evidence the oceans hold some potential surprises for us . . . that climatic changes can be sudden rather than gradual.

Basically, scientists are concerned over several potential positive feedback mechanisms. Warmer oceans could be less efficient absorbers of CO_2; they could release tons of methane (a greenhouse-effect trace gas) stored in sea floor mud; or they could, through evaporation, release more water vapor to the air. Water vapor is an even more efficient absorber of heat than carbon dioxide.

Richard Houghton and George Woodwell, researchers at Woods Hole Research Center in Massachusetts, are worried about something else. They fear global warming will accelerate the decay of land-based organic matter without changing the rate of photosynthesis. This, they say, would speed the release of both carbon dioxide and methane to the atmosphere, kindling even greater climatic warming.

In light of all the potential positive feedbacks, it is odd that critics of the predictions of greenhouse warming all seem to be lined up on one side, the side that says scientists have overdone it. Climate researchers themselves are the first to admit their models aren't perfect, that they are, according to Schneider, "dirty crystal balls." But it is important not to lose sight of one fact: that the models, because they haven't, for example, handled the atmosphere–ocean interaction very well, *could be just as apt to err in predicting too little atmospheric warming as too much.* That is, the greenhouse effect could be much worse than many scientists currently foresee. Even responsible critics concede this point. For instance, Robert Balling, an Arizona State University climatologist and greenhouse skeptic, cautions, "Although a one-degree warming is much more probable, it is ludicrous to say 5.5 degrees cannot happen."

As if to underscore the possibility of even greater environmental warming, a report in early 1989 ominously chronicled a more rapid rise in sea surface temperatures than scientists had suspected. The report, authored by Alan Strong, a National Oceanic and Atmospheric Administration (NOAA) research oceanographer, noted that between 1982 and 1988 weather satellites observed an increase in ocean temperatures of twice the magnitude previously estimated. Climate researchers found the satellite measurements interesting but were "a little nervous" about their validity. Still, the results can't be dismissed out of hand. Weather satellite observations, as opposed to conventional—and limited—water temperature measurements from ships, cover the entirety of the world's oceans between 60° N and 60° S. Conventional data are rife with gaps where no ships travel or take measurements. As the NOAA data are rechecked for errors and await corroborating observations, satellite monitoring will continue. If, during the 1990s, the more rapid oceanic warming is confirmed and proves to be more than transitory, then Broecker's concern ("the oceans hold some potential surprises for us") would be well-founded.

In addition to feedback problems, climate models also need improvement when it comes to simulating the effects of ozone in the upper atmosphere (recall the ozone loss-induced cooling discussed in chapter 6) and of "tropospheric anthropogenic aerosols." That is what scientists call tiny particles (aerosols) that result when man injects (anthropogenic) sulfur dioxide into the lower atmosphere (the troposphere), primarily by burning fossil fuels. These particles, actually tiny specks of sulfate—the culprits in acid-rain formation—are efficient sunlight reflectors. They may, in fact, play a significant role in slowing, but not stopping, greenhouse warming.

Ironically, in our efforts to clean up the atmosphere, we will likely reduce these global warming mitigators. The 1990 Clean Air Act calls for a significant cutback in U.S. sulfur dioxide emissions by 2000. This cutback will be achieved mainly by burning lower-sulfur coal and removing ("scrubbing") sulfur dioxide emissions from smokestack releases. Such procedures will not, however, reduce the amount of carbon dioxide being thrown into the atmosphere. It seems as though the net result will be cleaner air but greater global warming. (The Law of Unintended Consequences again.)

Finally, despite the shortcomings of climate models, they really aren't that bad. As John Mason, director general of the Meteorological Office, Bracknell, England, pointed out over a decade ago, "the models successfully simulate the major features of global atmospheric circulation and of present world climate, at least as far as the averaged conditions are concerned." Climate models also provide good simulations of the earth's seasonal changes, various historical climates (paleoclimates), Venus's warm greenhouse atmosphere, and Mars's cold conditions. Perhaps the most sophisticated climate model currently in use is the one at NASA. It is referred to in the research business as the GISS model . . . GISS for Goddard Institute for Space Studies.

The NASA Climate Model

In the early days of greenhouse-effect research, scientists used climate models to simulate current conditions, then instantly doubled the amount of CO_2 in the model atmosphere. The model was then run in a computer until climatic conditions stabilized at a new equilibrium. Maps could then be generated showing the changes in such elements as temperature, precipitation, soil moisture, and cloudiness that might result from doubled atmospheric carbon dioxide. Since this represented a large jump from one atmospheric state to another covering a period of decades, researchers weren't really shown what climates in between might be like; that is, the model didn't indicate climatic conditions intermediate to the current state and the one with doubled CO_2. By using analogs, such as the 1930s for the mid- to late 1990s, climatologists could make some good guesses, but they weren't helped out by the computer models, at least until NASA's GISS model was born. (Analogs, by the way, will still be useful, since—for various stops along the way to a warmer world—

climatic analogs can be compared with appropriate climate model outputs.)

The NASA climate model, the one on which James Hansen based his 1988 Senate panel testimony (see chapter 1), marks a significant advance in climate simulation. The model naturally employs all the laws (equations) of physics mentioned earlier in this chapter. The model, any model for that matter, solves the equations for a series of "boxes" into which the model atmosphere is divided. The boxes are all mathematically and logically linked to one another. In the NASA model the boxes are quite large; horizontally, each is only slightly smaller than the states of Utah, Colorado, Arizona, and New Mexico combined; vertically, the boxes are divided into nine (atmospheric) layers. This may be contrasted with the models used to forecast weather on a short-term, day-to-day basis . . . the models TV weathercasters look at. These models typically use boxes that are measured in tens of miles or kilometers horizontally (perhaps the size of a county), not hundreds of miles or kilometers. (Thus you can see that climate models, because of the "coarse" grid they use, can't depict regional differences in great detail. Analogs are sometimes handy in helping out with this problem.)

The NASA model does other nifty things that climate researchers like: it computes cloud amounts and heights, models evaporation, includes the seasonal and daily distribution of solar heating, calculates soil moisture and surface albedo (reflectivity) based on local vegetation, computes snow depth and albedo, and models convection. Convection is the vertical movement of heat through the atmosphere; among other things, convection produces rain showers and thunderstorms.

But what really distinguishes the NASA model from others* is its ability to "inject" CO_2 and trace gases into the model atmosphere in a realistic, time-phased fashion, rather than just doubling the amount of atmospheric CO_2 and letting the model run from there. Thus the NASA model is able to provide us pictures, or maps, of what our climate may be like in the 1990s, the 2000s, the 2010s, and so on.

Unfortunately, even in the NASA model the simulation of interactions between the atmosphere and the oceans remains primitive. This has led researchers to warn: "In the real world, climate changes at the ocean

---

*Other climate models commonly used to study the greenhouse effect include ones from the National Center for Atmospheric Research (NCAR), Princeton University's Geophysical Fluid Dynamics Laboratory (GFDL), Oregon State University (OSU), and the British Meteorological Office (BMO).

surface may induce changes in ocean heat transports, thus leading to other, perhaps larger, climate changes." Recall the positive feedback problem mentioned earlier in this chapter.

Before running any experiments with the NASA model, researchers made certain it didn't have any built-in biases. Using the atmospheric composition and climatic conditions that existed in 1958 as a starting point because that was when we first had reliable measurements of atmospheric CO_2, scientists allowed the model to run for a simulated 100 years. No carbon dioxide was added to the model atmosphere, and no other external influences (such as volcanoes or changes in the sun's output) were allowed to force climatic changes.

At the end of its simulated century run, the model atmosphere had virtually the same temperature it did initially. Along the way, however, it experienced upward and downward temperature trends much as the real atmosphere does. These warming and cooling cycles showed only slightly less variability than has actually occurred over the past 100 years, leading scientists to feel they had a very good model.

Playing Games

The beauty of such models is that researchers can play "what if" games. They can vary the rate at which CO_2 and trace gases are allowed to enter the model atmosphere, and do other things such as simulate volcanic eruptions. Volcanoes often inject huge amounts of sulfur dioxide into the atmosphere. This ultimately results in a haze of sulfuric acid droplets (volcanic aerosols) in the high atmosphere. These high-altitude volcanic aerosols reflect sunlight, thus inhibiting warming or initiating short-term cooling trends.

Using the NASA model, researchers developed three greenhouse-effect scenarios, A, B, and C. Scenario A assumes continued growth of atmospheric CO_2 and trace gases at rates compounded annually. Growth is thus exponential, meaning the concentration of greenhouse gases increases by a larger amount each year. In this scenario the growth rate of CO_2 is set at 1.5 percent per year, which is roughly what it was during the late 1980s oil glut. Before the glut and after 1970, the rate was about 1 percent. Trace gas concentrations in scenario A are allowed to increase at rates typical of the 1970s and 1980s. The scenario also models several hypothetical or crudely estimated trace gas trends (for ozone, high-altitude water vapor, and minor chlorine and fluorine compounds) that are not included in the other scenarios. NASA scientists refer to scenario A as "business as usual."

In scenario B ("limited emissions"), a linear as opposed to an exponential growth of greenhouse gases is assumed. This means that growth rates decrease slightly each year, implying a reduction in per capita emissions. Growth still occurs, though, because the earth's population continues to swell. Researchers took full advantage of the model's "what if" capabilities in this scenario by inserting some large volcanic eruptions for the years 1995, 2015, and 2025. These are just arbitrary choices, of course, and don't imply that anyone expects big volcanoes to erupt in those particular years. But the researchers wanted to see how significant volcanic eruptions might affect overall atmospheric temperature trends. Thanks to Mount Pinatubo, however, a real-world test is currently underway.

Finally, in scenario C, drastic cuts in the use of fossil fuels are assumed between 1990 and 2000, and CFC emissions are assumed terminated by 2000. The net result is that the greenhouse-gas growth rate is zero by the turn of the century. Since this represents a more drastic curtailment of emissions than has generally been imagined, scenario C has been subtitled "Draconian emission cuts." Like scenario B, scenario C includes volcanoes for 1995, 2015, and 2025.

For the long haul, NASA researchers feel scenario B to be the most likely of the scenarios. This is because scenario C assumes more drastic cuts in fossil fuel usage by the year 2000 than can realistically be expected. Scenario A, on the other hand, "must eventually be on the high side of reality in view of finite resource constraints and environmental concerns."

Still, for the near term, scenario A may not be far off base. Since no significant steps have yet been taken to curtail CO_2 emissions on a global basis, Hansen believes that scenario A is certainly "plausible" for the 1990s. He reminds us, however, in light of Mount Pinatubo's eruption, that no large volcanoes have been modeled in scenario A. Hansen also says he expects CFC emissions to diminish because of the Montreal Protocol (see chapter 6) *if all the signatory nations abide by the accord.* All things considered, though, one gets the feeling that even through the early decades of the twenty-first century, scenario A isn't exactly out the window.

Besides the modeling uncertainties of such things as volcanoes and ocean–atmosphere interactions, there is one additional uncertainty worthy of note. It's called solar irradiance. Solar irradiance refers to how

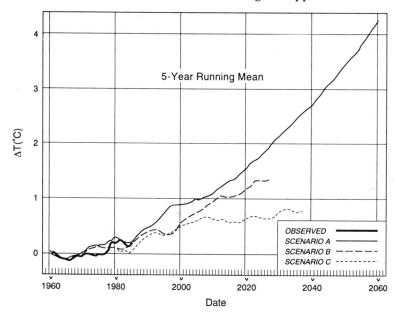

FIGURE 7.1 Global temperature forecasts by the NASA model: three scenarios. *Source:* J. Hansen, I. Fung, A. Lacis, D. Rind, S. Lebedeff, R. Ruedy, and G. Russell, "Global Climate Changes as Forecast by Goddard Institute for Space Studies Three-Dimensional Model," *Journal of Geophysical Research,* vol. 93, pp. 9341–64, fig. 3(b), copyright 1988 by the American Geophysical Union.

much energy the sun puts out. Such energy varies with time, leading to fluctuations in global temperature. For example, solar irradiance decreased between 1979 and 1985 to the extent that any significant warming due to the greenhouse effect in that period probably was negated.

Recent data suggest that solar irradiance has since bottomed out. Because changes in the strength of the sun's output are not modeled in the various scenarios, it's possible that an increase in irradiance in the near future could enhance predicted greenhouse warming. That's only in the near future, however. In the longer term—say, after several decades—trends in solar output one way or the other (and trends in the frequency of volcanic eruptions, too) wouldn't make any difference; climate changes caused by the ever-increasing greenhouse gases would overwhelm everything else.

Forecasts for a Warmer World

Figure 7.1 depicts the global temperatures forecast by the NASA model

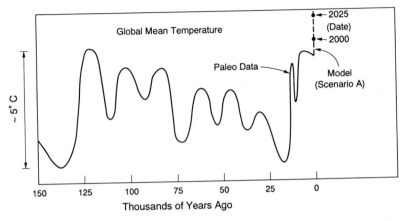

FIGURE 7.2 The Global temperature trend over the past 150,000 years (with projected greenhouse warming added). Reproduced, with permission, from J. Hansen, et al. "Prediction of Near-Term Climate Evolution: What Can We Tell Decision-Makers Now?" *Preparing for Climate Change: Proceedings of the First North American Conference on Preparing for Climate Changes*, 1987, pp. 35–47, fig. 2.

for the three scenarios. Also shown is the observed, or actual, global temperature curve beginning in 1958. The curves are plotted using five-year running means, which are five-year averages moved one year ahead for each year plotted. For instance, the first point shown on the graph is for 1960. It's actually the average of temperatures for 1958, 1959, 1960, 1961, and 1962. Five-year running means suppress year-to-year variations in temperature, resulting in smoother curves than does plotting each year individually. Figure 7.1 also shows that global temperature trends predicted by the NASA model replicate very nicely the observed trends since 1958.

The figure further shows that there is relatively little difference (0.4° C, or 0.7° F) in forecast global temperatures by the year 2000 regardless of which scenario is followed. After that, however, the temperature curves depart significantly from one another, with the larger difference manifest between scenarios B and C than between A and B. By 2020, global temperatures in response to both scenarios A and B are predicted to exceed anything the earth has experienced in the past 150,000 years!

This trend is shown schematically in figure 7.2 for scenario A. Note the extreme rapidity with which temperature changes are forecast to occur. Changes that previously took place over the course of centuries—or more usually thousands of years—would occur in just a few decades. The implications of such extremely rapid changes are discussed in chapter 8.

By the beginning of the second decade of next century, global warming under scenario A is forecast to reach 1.5° C (2.7° F), while under scenario B the figure would be closer to 1.1° C (2° F). With scenario C the warming would be "limited" to 0.6° C (1.1° F) and would be in the process of leveling off, but keep in mind that C is by far the most unlikely of the scenarios. I'll go so far as to say it just won't happen.

By 2030 the scenario A temperature—reflecting a warming of a little over 2° C (3.6° F)—is predicted to be in the lower part of the range generally accepted as the likely response to doubled atmospheric CO_2. Under scenario B this level of warming would not be reached until around 2060. (Remember, all the scenarios consider more than just CO_2; they include various greenhouse trace gases as well.)

After 2000, while the scenario C curve flattens out and the scenario B trend becomes linear, the scenario A plot continues to increase exponentially. By 2060 the scenario A global temperature is predicted to be in excess of 4° C (7° F) and still accelerating.

Although NASA researchers have deemed scenario B the most likely in the long run, it seems unlikely—given our recent inaction in curtailing CO_2 emissions (see chapter 6)—that we will shift to a scenario B–type world within the next decade or two. In reality, assuming we finally do *something* in response to greenhouse warming, we will probably heat our planet, by the middle of next century, to somewhere *between* what scenarios A and B predict. This is not necessarily comforting. Remember, even under scenario B the warming continues, albeit at a linear rate instead of scenario A's exponential rate.

Mount Pinatubo

Large volcanic eruptions have historically produced noticeable climatic effects. Tambora, on the island of Sumbawa in Java, erupted in 1815. The explosion supposedly produced the greatest ejection of volcanic material in history as it blew away 4,000 feet (over a kilometer) of the mountain's height. Many researchers think Tambora's volcanic haze was a key contributor to the "year without a summer" in 1816. Unusually cold and inclement weather plagued the entire Northern Hemisphere that summer. In New England, snow whitened parts of the countryside in June, light frosts struck in July and August, and heavier frosts wiped out the corn crop in September.

In 1883, an Indonesian volcano, Krakatau, blew up with such force that it produced a shock wave that traveled seven times around the world. The sound of the explosion itself was heard 3,000 miles (4,800 km) away and is generally considered to be the loudest noise in the history of man. The climatic legacy of Krakatau followed in 1884: globally, it was the coolest year on record in the past 110-years-plus.

Mount Pinatubo's 1991 eruption spewed the greatest amount of volcanic aerosols into the atmosphere since Krakatau. The initial effects were a series of colorful sunsets initiated by the volcanic haze—some fifteen to sixteen miles (25 km) high—as it spread over the Northern Hemisphere in the summer and fall of 1991.

The NASA climate model was run using data describing Mount Pinatubo's volcanic haze, and the result was predicted global cooling, overriding greenhouse warming, during 1992 and 1993. Using both scenario A and scenario B (with no additional volcanoes), the model also forecast a return to record global warmth after 1995.

How well the model handles the predicted cooling—since it is forecast to be significant, albeit transitory—will give researchers excellent insight into how good the NASA model really is. Assuming the measurements of Pinatubo's gas and dust were correct, Hansen said the volcano will provide an acid test for global climate models. In an unusually candid statement, he went on to say, "If we do not observe substantial measurable global cooling in 1992 and 1993, the climate models are simply wrong."*

The flip side to that statement is that if the forecast cooling materializes, we can put even more credence in the models' predictions of global greenhouse warming.

Pictures of a Warmer World

One of the strengths of the NASA model is its ability to create pictures, or maps, of what our climate may be like at any particular time in the future as we inexorably warm the earth. For example, the model can generate a map of temperature increases relative to current temperatures for (a) any particular decade in the future (e.g., the 2010s), (b) any particular season

---

*By mid-1992 National Oceanic and Atmospheric Administration scientists, using satellite data, had detected about a 1° F (0.6° C) drop in global temperatures.

of any decade in the future (e.g., summer 2010s), or (c) any particular month of any particular year in the future (e.g., July 2012).

These maps, as well as those from other models, will be referenced in the following chapters on longer-range greenhouse-effect climate scenarios (chapter 8) and regional consequences (chapters 9 through 11).

In general, the NASA model shows a greater degree of warming taking place over continental areas than over oceans and more warming occurring in high latitudes (polar regions) than in low latitudes (tropical regions). Through the latter part of the 1990s, NASA scientists note, the model suggests greater-than-average warming in the southeastern and central United States and relatively cooler conditions or less-than-average warming in the western United States and much of Europe.

Shortly after the 1990s, however, there will likely be no "cooling" areas left in our world.

Skeptics

Skeptics will take exception to the preceding statement . . . and a lot of other statements relative to the greenhouse effect. Certainly there are a number of scientists who don't believe a greenhouse effect exists at all, and have said so. But that's all they've done: said so. No researcher has yet used a climate model to suggest we're doing anything other than warming our world with the release of CO_2 and certain trace gases into the atmosphere.

But skeptics point out that even using models we have problems predicting the weather just a few days in advance; thus, they say, it's absurd to think we can do it a few decades in advance. The problem with that reasoning is that researchers are *not* predicting the weather a few decades in advance. They're predicting the climate. In the long range, climate is easier to forecast than weather, since climate patterns result from precipitation and temperature figures averaged over long periods. The hour-to-hour or day-to-day movement of weather systems is not of concern in climate prediction. What is of concern is the resultant, or mean, positions of those weather systems over a period of decades. Forecasts of weather for specific days in the future are, indeed, not possible. Forecasts of climatic conditions for particular decades in the future are. Remember, the global warming measured since 1958 is equal to the amount the NASA climate model computed.

Critics have also argued that the greenhouse effect may be self-defeating. Specifically, they theorize, as global temperatures climb, more evaporation of moisture from the surface of the earth will take place. More evaporation will lead to more cloudiness. More cloudiness will result in more sunlight being reflected away from the earth. And this, in turn, will cause global temperatures to fall or at least level off. Gates, from Oregon State University, retorts that this "doesn't make much meteorological sense."

The earth's weather develops in coherent patterns, such that if parts of the globe grow cloudier and wetter, others turn sunnier and drier. This has to do with the way the westerlies shift with changing climate, as was discussed in chapter 5. A few places—India and parts of Australia and South America, for instance—*will* likely be cloudier and damper in a warming world. Climate models as well as analogs agree on this. But climate models are virtually unanimous in predicting that large areas of the earth, particularly midlatitude midcontinental regions, will become hotter and drier in a greenhouse world.

Hansen and other NASA researchers have taken a close look at this. Under their scenario A they determined that drought conditions, which traditionally parch about 5 percent of the earth's land area in any given year, would expand significantly as our world warms. By the late 1990s drought would stalk roughly 10 percent of the land; by the 2020s the figure jumps to 25 percent and by midcentury to a frightening 45 to 50 percent!

Even for regions likely to turn wetter the news isn't necessarily good. Almost all of the increased rainfall would come in the form of heavier showers or thunderstorms—cloudbursts—as opposed to widespread, long-lasting light precipitation. This means more of the rainfall would run off instead of being soaked up by the ground. This isn't beneficial, of course, and indeed may lead to greater flooding.

Another counter-greenhouse-effect theory is that the atmosphere is "self-cleansing" when it comes to heat. The argument goes that tropical thunderstorms and cyclones will punch through the greenhouse-gas blanket vertically and blow away excess heat. Gates notes there is no research supporting this view. Hansen is even blunter in his criticism of the theory; it's "crackpot," he says.

Some meteorologists feel our recent global warming can be attributed more to the development of a warm tropical ocean current called El Niño than to the greenhouse effect. El Niño is the name given a periodic general warming of the tropical eastern Pacific Ocean. El Niño was present during 1987, for example, and researchers acknowledge that it was at least partially responsible for the atmospheric warming that occurred then. They also acknowledge that in the absence of El Niño, global temperatures may briefly level off or even decline slightly (but after the '87 El Niño this did *not* happen). They warn, however, that such an interruption of atmospheric warming would be only temporary. El Niño returned in 1991–92 and was in part responsible for the devastating floods in southern California in February 1992.

One of the most vehement critics of the greenhouse theory—and of scientists and authors who support it—is Pete Leavitt, an executive with Weather Services Corporation of Bedford, Massachusetts. It's all "horseshit," he says, referring to predictions of impressive global warming. Leavitt acknowledges there has been an upward trend in worldwide temperatures, but he dismisses it as minor and within the range of natural climatic fluctuations. Without offering quantitative evidence, Leavitt says the data supporting a more significant warming are unreliable. But researchers disagree sharply, indicating they have worked diligently to remove biases from global temperature records and that their work is open to review.

Leavitt carries his polemic further, railing against climate models as lacking credibility ("they don't know how to handle clouds"—something climate modelers freely admit, as noted earlier in this chapter) and against researchers supporting the greenhouse theory as charlatans. Leavitt, taking a page from "The Emperor's New Clothes," seethes, "Somebody should point out these guys are naked." Leavitt sees the dire forecasts of greenhouse warming as a ploy to continue receiving government grants, the money that keeps research organizations afloat.

Hansen retorts: "This is the most far-fetched claim I have ever heard. Most researchers are aware of the funding difficulties we [at NASA] have faced since 1981, and they are not anxious to get in the same boat. Speaking out may get attention for the field, but I can guarantee that it doesn't help get you a research grant."

In 1981 Hansen and six other atmospheric physicists at NASA authored a research paper "warning" of CO_2-induced warming. Since this was during a time that many scientists thought the earth was cooling off, the "warning" made the front page of the *New York Times*. The article attracted a great deal of attention and greatly displeased the U.S. Department of Energy (DOE), the lead agency for CO_2 greenhouse-effect research. Hansen says the head of the DOE program, Fred Koomanoff, "made it clear that he did not like our paper or the publicity that surrounded it." Since then DOE has turned down all NASA requests for support of greenhouse-effect research* (The NASA research, however, has been supported by grants from the EPA.)

Finally, there is the argument that the return of an Ice Age may save us. Researchers have discovered that because of the earth's orbital eccentricity and the tendency of the earth's axis toward precession and tilting back and forth, the amount of solar energy reaching the earth varies over periods of thousands of years. It is true that we probably are heading back toward an Ice Age. It won't save us from the greenhouse effect, though; it will be a little late. About 20,000 years late. The greenhouse effect is here now. It will be another 200 centuries before the earth can wobble and tilt its way back to a position that would permit the great sheets of ice to return to the Northern Hemisphere.

No, an Ice Age won't save us from a greenhouse world. Only we can do that. What I don't have in mind, however, is the "Geritol solution."

The Geritol Solution

The "Geritol solution" came to light in mid-1990. Scientists, searching for a way to actively counter global warming, announced a scheme to enlist marine algae in the battle. The plan, simplified, is this: by dumping hundreds of thousands of tons of iron (thus, the "Geritol solution") into the world's oceans, the growth of tiny algae called phytoplankton could be stimulated. The little algae, like all plants that carry out photosynthesis, consume carbon dioxide. So if—through iron "fertilization"—the proliferation of phytoplankton could be encouraged, they should soak up much of the excess CO_2 we've been injecting into the atmosphere. That, in turn, would help stem or reverse global warming. "You give me half a tanker full of iron, I'll give you another ice age," says John Martin (tongue-in-cheek) of Moss Landing Marine Laboratories in California.

While the idea seems technically feasible, success with weather modification on much smaller scales has, at best, been limited. Results of experiments to make rain, modify hurricanes, and suppress hail have

---

*This "ostrich approach" to the greenhouse effect continued with the Bush Administration. In mid-1989 the White House Office of Management and Budget (OMB) changed the text—over the protests of Hansen—of NASA testimony on global warming, delivered to the Senate Subcommittee on Science, Technology, and Space. The OMB changes were designed to make Hansen's conclusions about the greenhouse effect seem less serious and more uncertain than he intended. Earlier in the year the OMB had attempted unsuccessfully to edit similar testimony provided by Jerry Mahlman, director of NOAA's GFDL.

often been described as tentative, inconclusive, or lacking evidence of changes. So now scientists want to deliberately modify the global climate? The idea is a nonstarter . . . probably less from a technical standpoint than from a political one, however.

Consider, for example, rainmaking. If experimenters *are* able to enhance rain in one area, at whose expense should that enhancement come? The next county or state downwind? A neighboring country? (If increased amounts of moisture are precipitated out of a cloud at a particular location, there is less left for other locations.) On the other end of the scale, if flooding rains should fall in a region of weather-modification work, how could it be proved the modification efforts didn't cause or contribute to the flood? While scientists might be able to satisfy themselves that such was not their doing, to the guy whose house is three feet deep in mud, or whose automobile became a submarine, the proof would never be conclusive.

A case in point: during one of the early efforts (in 1947) to diminish the intensity of hurricanes by seeding them with silver iodide, a storm that had been moving northeastward off the Georgia coast took a sudden turn westward and smashed inland near Savannah. In fact, the seeding probably had nothing to do with this abrupt change of course. But try telling that to a family whose coastal home has become kindling.

Remember, too, the Law of Unintended Consequences. A classic real-life example of that canon occurred during the early days of the Vietnam War. Military experts decided that a quick-and-easy way to defoliate the jungle would be to simply burn it. With the help of lots of napalm the air force set out to do exactly that in a test run outside Saigon (now Ho Chi Minh City). Sure enough, they got a big fire going. But they had forgotten to calculate how Vietnam's warm, moist, unstable tropical atmosphere would react to the sudden addition of large amounts of heat. It did exactly what any respectable atmosphere would do. It responded to the laws of thermodynamics and built a large thunderstorm directly over the fire. A tropical cloudburst ensued, and within minutes the fire was gone and the jungle was still there.

Given (a) that we cannot be 100 percent certain of where a greenhouse world will take us, (b) that we can never be sure of the consequences, both climatic and biologic, both intended and unintended, of something like the Geritol solution, (c) that a few countries might, in fact, welcome greenhouse-effect climatic changes, and (d) that the international community cannot even get together to take actions with *predictable* outcomes to stem global warming, then it seems virtually certain the nations of the world will never agree to a hypothetical approach to "fixing" things. Let us not forget, either, that carbon dioxide is only part of the greenhouse problem.

We'd like an easy way out of our environmental mess, of course. But armies of little plants gobbling up CO_2 aren't the answer. Neither is anything else. As Sherwood Rowland of the University of California at Irvine said in late 1991, "Nothing proposed yet is even remotely feasible." Rowland and some of his colleagues expressed concern that people are living under the misconception that when problems become urgent, scientists will solve them. "That isn't going to happen," he warned.

No, technology and science aren't the answer. People are. You and me. You and me and the actions we take or don't take are going to determine our environmental future.

8 / Crocodiles in New York

In the late 1970s Arthur Herzog wrote a novel about the greenhouse effect, *Heat*. The book's protagonist is Lawrence Pick, an engineer who warned of the dire consequences of a global temperature rise. No one would listen, of course, and a nightmare of altered weather patterns ensued. As runaway atmospheric warming took over, Pick reflected on where it all might end. "He . . . let his mind run. It came to New York years from then if the heat rise continued. Greatly shrunken by the rising sea, the city would resemble a huge Mayan ruin, with tall buildings covered by creepers and moss. Perhaps crocodiles would float in the lakes and reservoirs. . . ."

Is this where it all might really end? Crocodiles and Spanish moss in Manhattan? Probably not. But if at all, not for *thousands* of years from now. In the more immediate longer range—say, over the next fifty to 100 years—there are, however, some serious if not quite so dramatic greenhouse-effect problems with which we will have to deal. There are threats that will be manifest within a decade or so—such as heat and drought in the midsection of the United States—and others that will not be readily apparent until the middle of the twenty-first century—such as a rising sea level.

This chapter peers into the next century and presents a global overview of the menace of a greenhouse world. The three chapters following this one focus on the United States and tackle the more difficult challenge of outlining some regional consequences of a hotter world. There is general agreement on where the greenhouse effect is taking us globally: a warmer Earth. There is even general agreement on the consequences some countries will face: a hotter and drier middle America, for instance. But when it comes to pinning down specific consequences for particular regions of a country the uncertainties mount . . . remember the very large grids used by the climate models. Still, that doesn't mean the threats should be ignored. They are serious. And whether we want to think about them or not, some will change our way of life.

Don't Write Off Florida Yet

A favorite greenhouse-effect topic of the news media is the threat of rising seas resulting from the world's melting ice and snow. South Florida, for instance, is pictured as becoming the Florida Shoals, Orlando as meta-morphosing into a seashore resort. This won't happen, at least not in our lifetimes or in those of our grandchildren. I have to think we won't ever let it happen.

But the sea level *is* rising, and it's going to continue to rise . . . even more quickly than it has. For some parts of the earth this will present an almost insurmountable set of problems. And unlike the submergence of Florida, they're problems that will occur within our lifetimes, or certainly within the lifetimes of our grandchildren.

Over the past century the sea level has crept upward around half a foot (16 cm), mostly in response to the global warming of about 1° F (0.6° C). Indications are that since 1930, however, the sea level has been coming up more rapidly, at a rate of around nine inches (23 cm) per century. Even this rate is slow compared with what's in our future.

In a greenhouse world the sea level will rise for several reasons. First, there is "thermal expansion" of water: water expands as it grows warmer. Next, there is runoff from melting continental ice caps, primarily those of Greenland and Antarctica. (Arctic ice is mostly frozen sea water, so its thawing won't contribute significantly to an increase in sea level.) Finally, there is runoff from the melting snow and ice of Alpine glaciers. Scientists freely admit they don't fully understand the details of all these processes, so there is a wide range of estimates regarding how fast and how far we can expect the oceans to elevate.

While some estimates suggest as much as a five-foot (1.5-m) increase in sea level before midcentury, an average of several estimates puts the figure closer to two or three feet (1 meter or less). If we don't curtail our use of fossil fuels and put the brakes on the greenhouse effect within the new few decades, predictions are we will commit the earth to an ocean rise of up to twelve feet (3.7 m) by 2100; a mean figure, however, sets the increase nearer five feet (1.5 m). (The disintegration of the West Antarctic ice sheet could add sixteen to twenty feet [5 to 6 m] to the sea level, but such a "collapse," if it happens at all, is centuries away.)

Let's, for a moment, consider a three-foot rise in sea level within the next fifty years. Three feet. Hardly enough to envision Miami as becoming an Atlantis, or Atlanta as becoming a Miami. Thirty-six inches . . . ninety-one centimeters . . . not much on a calm day at the beach. But calm days aren't what we have to worry about. Let's think about three feet on stormy days, especially stormy days along the flat littorals of the Atlantic and Gulf coasts.

Picture a beach house, perhaps one that in the past has proved vulnerable only to the ravages of a rare category four hurricane, a storm that generates tidal surges thirteen to eighteen feet (4 to 5.5 m) above normal. With a three-foot rise in sea level, the beach house would be threatened by category three hurricanes, storms that push up surges nine to twelve feet (2.7 to 3.7 m) above normal, and storms that are currently three times more common than category fours! But that's not the only problem. If you'll recall from reading chapter 3, our greenhouse world with its warmer oceans will, in all likelihood, foster tropical cyclones of even greater ferocity than in the past. Or, saying it more directly, the number of category three, four, and five hurricanes will increase—a potential deadly complement to a rising sea level.

There will be other problems, too. Ones not limited to the United States. Worldwide, the rising oceans will enhance shoreline erosion; inundate coastlines, beaches, and wetlands; disrupt coastal industries; allow salt water to penetrate inland lakes and ponds, affecting drinking water and ecosystems; and permit saline intrusions into coastal freshwater aquifers. In some parts of the world the problems will be less insidious.

Bangladesh—Death in the Delta

In a greenhouse world with rising seas, it is perhaps not overstatement to warn that a nation like Bangladesh will teeter on the edge of a holocaust . . . one of water, not fire.

Bangladesh is a tiny nation, slightly smaller than the state of Georgia, strangled by a population of almost 120 million (with 7,000 new residents born each day). By way of contrast, Georgia is home for about six-and-a-half million people. Bangladesh is dominated by a low-lying, flat delta, the Ganges-Brahmaputra-Meghna Delta on the Bay of Bengal. In this delta region thousands of islands called chars exist, products of tons of silt spilling from the mighty rivers. The silt shifts continuously, tearing down some chars while at the same time building new ones. The chars are home to literally millions of farmers attempting to subsist on the rich, silty soil. In all, over thirteen million Bangladeshis struggle to survive on chars and other delta ground less than ten feet (3 m) above sea level. And unlike the situation in the Netherlands, no dikes protect the Bangladesh lowlands.

As if this weren't a big problem in itself, the coastline where the delta meets the Bay of Bengal is concave and the coastal water is shallow. A better setup for generating massive storm surges from tropical cyclones could not be imagined. When a cyclone (which is what tropical storms and hurricanes are called in that part of the world) slashes in from the Bay of Bengal, the results can be numbingly tragic. In November 1970, a cyclone pushed a twenty-foot (6-m) storm surge over the chars and up the delta; 300,000 people died.

After the 1970 disaster, the government built hundreds of multistory concrete shelters on the more stable delta islands. In addition it installed warning sirens, distributed radios, built up embankments, and planted mangrove forests. Still, the dying continues. As recently as April 1991, over 140,000 lives were claimed by a cyclone. Many chars were completely inundated; tens of thousands of people, especially children, never made it to the shelters.

The pathetic plight of this impoverished country will grow even more desperate in a greenhouse world. The population will continue to burgeon; the rising sea, even in good weather, will continue to claim delta islands; and the Bay of Bengal will continue to bear death-dealing cyclones, some of even greater intensity than has been known. Remember the 800-millibar supercanes warned of for the northwest Gulf of Mexico (chapter 3)? The only other place in the world where storms of such magnitude may have a potential of occurring is the northern Bay of Bengal, near the coast of Bangladesh.

So what do we do with thirteen—or fourteen or fifteen—million people threatened with annihilation? "When the flooding starts," asks Gina Maranto, staff writer for *Discover* magazine, "do you suppose they'll simply pack up in an orderly fashion and be politely allowed to file into India or Burma?"

You probably know the answer to that. And Bangladesh isn't the only place that such a dilemma will be faced. Low-lying coral-island nations like the Marshalls in the western Pacific, the Maldives off the west coast of India, and a few tiny places in the Caribbean also face extinction as the ice caps and glaciers drip and the ocean waters rise. As Robert F. Van Lierop, president of the Alliance of Small Island States, laments, "We do not have the luxury of waiting for conclusive proof of global warming. The proof, we fear, will kill us."

Thirteen or fourteen or fifteen million people. I know this doesn't seem to be a problem on *our* back doorstep, but it should stir our conscience, and it should silence the critics who suggest that a greenhouse world won't be so bad . . . that we can learn to live with such a thing. Tell that to the Bangladeshis.

It is true that in the United States we will not face death and destruction on a massive scale, and I suppose that robs some urgency from the headlines. But what might the headlines be? Maranto asked the same question in the January 1986 *Discover* article in which she addressed the Bangladesh dilemma. "What will the headlines be in 2035? GREAT PLAINS DROUGHT IN EIGHTH YEAR; GLOBAL FOOD CRISIS DEEPENS . . . ?"

Another writer in hyperbolic orbit? Let's see. Let's go back to climate models and climate analogs for a moment.

Three or Four Degrees Doesn't Sound Like Much

Virtually all climate models predict that a doubling of atmospheric CO_2 will lead to a warmer Earth, somewhere in the range of 3° to 8° F (1.5° to 4.5° C). The sophisticated NASA model forecasts a global temperature rise of over 3.6° F (2° C) by the middle of next century . . . 2030 for scenario A, 2060 for scenario B (chapter 7).

Three or four degrees doesn't sound like much. After all, humans probably aren't sensitive to temperature changes of any less than three degrees. But that's not the point. Remember, it's not a matter of just taking three or four degrees and tacking that on to the mean temperature of every place in the world. Things aren't that simple.

What's important is the way patterns of temperature and precipitation will change regionally, and the degree to which the frequency and intensity of certain severe weather events will be altered. At least for the United States in the 1990s, we've already considered how the frequency and severity of such phenomena as heat waves, droughts, and hurricanes will likely escalate.

Twenty-first-Century Temperatures

Let's now examine what some of the models are telling researchers about the way large-scale regional weather patterns might well change. The different models all agree that polar regions will undergo the greatest warming in a greenhouse world. For instance, the Oregon State University (OSU) climate model predicts Northern Hemisphere polar regions will warm in excess of 14° F (8° C) as a response to globally doubled CO_2.

This large polar warming will come about for a couple of reasons. The first concerns water vapor. Water vapor normally absorbs a great deal of the earth's heat radiation. Thus, in areas where water vapor is fairly abundant, such as the middle latitudes, CO_2 competes with water vapor for absorption rights. But in polar regions where the air is quite cold and therefore cannot hold very much water vapor, CO_2 has the heat radiation pretty much to itself. So the addition of CO_2 to areas of the earth that are quite dry—such as the poles (and deserts)—produces a relatively large increase in heat absorption and, therefore, warming. In places where there is a lot of water vapor, however—such as over tropical oceans—the absorption and warming effects are noticeably smaller.

The second reason follows the first. Once the warming in the polar regions is under way and ice and snow begin to disappear (melt), the warming accelerates. This is because as the ice and snow melt, exposing land and water, surfaces that previously reflected much of the sun's energy back to space become radiation "absorbers" instead, encouraging further warming. The whole process is self-enhancing (a positive feedback . . . chapter 7): more warming leads to more ice melting, which leads to more solar radiation being absorbed, which leads to more warming, etc. (A recent study using satellite data determined that the extent of ice in the Arctic Ocean declined by about 2 percent from 1978 to 1987. Could this be an early sign of greenhouse warming?)

There is also general agreement among the various climate models that in a high-CO_2 world, midlatitude, midcontinental areas will undergo a greater degree of warming than other regions outside the poles. This is because continents warm (and cool) more rapidly than oceans and because midcontinents tend to be relatively dry, thus allowing CO_2 to be a more effective heat absorber there than elsewhere (see two paragraphs preceding). For example, the OSU computer model suggests that with doubled CO_2, mean annual temperatures in the United States from the Great Lakes to northern Alabama and Georgia will be more than 7° F (4° C) higher than current averages. Similar temperature increases are forecast over a broad area of western Asia into the Russian steppes and the southwest Asian deserts of Iran, Afghanistan, Pakistan, and northwest India.

A climate model used at NOAA's Geophysical Fluid Dynamics Laboratory (GFDL) at Princeton University appears to support the OSU predictions for these regions. With CO_2 doubled, the GFDL model predicts summertime warming in excess of 11° F (6° C) for parts of the Russian steppes and much hotter summers for all of North America. In North America, the greatest increase in heat is forecast over the plains of the northern United States and southern Canada, where summer means could soar to over 14° F (8° C) above current normals! In terms other than statistical, the GFDL model is warning that summers in North Dakota and southern Manitoba will become more like the present-day sizzlers that fry northern Texas and southern Oklahoma.

Twenty-first-Century Precipitation

Predicting how precipitation regimes will be altered in a greenhouse world is a bit more challenging than forecasting temperature changes. This is because the processes that produce precipitation are more complex than the ones controlling temperature and are therefore more difficult to model. Syukuro Manabe and Richard Wetherald of the Geophysical Fluid Dynamics Laboratory have expended a great deal of effort in this area, however. Their GFDL model is now one of the most widely referenced when it comes to examining precipitation patterns in a world with doubled CO_2.

Actually, the model's output is not precipitation but soil moisture, which is really what agriculturalists and hydrologists are most concerned with. Hand in hand with the large increase in summertime temperatures forecast for North America, Europe, and Asia, the model predicts significantly drier soils. The most severe drying (a 50 percent or greater loss of soil moisture) is foreseen for the Great Plains of the northern United States and southern Canada, parts of southwest Europe (France and Spain), and the deserts of southwest Asia north of 30° N. Drying of a similar magnitude is forecast for extreme northwestern Africa.

The model doesn't predict only drier conditions. Significantly in-creased summer soil moisture is forecast for the Indian subcontinent, and slightly wetter soils are suggested for much of Australia, particularly in the winter.

Climatic analogs back up some of these climate model predictions, and where there appears to be an agreement between what the computer models and the analogs are telling us, we can feel much more certain that the climatic changes foreseen really will happen.

Our Once and Future Climate

Within the next decade or two global temperatures will likely reach levels that prevailed about 4,000 to 8,000 years ago. That was an era called the post-Glacial Optimum, or Altithermal Period. From exhaustive studies of marine and animal fossils, vegetation and bog growth patterns, tree growth-rings, ice cores, and pollen buried in sediments—climatic records etched in nature's logbook—researchers have a rough idea of what precipitation patterns looked like during the Altithermal Period, although the patterns did not necessarily occur simultaneously. (Remember, the period covered 4,000 years.)

The Altithermal analog seems to agree with the GFDL soil-moisture model in several regions: the United States, India, and Australia. During the Altithermal Period, evidence suggests, India and western Australia were wetter and the United States from the interior West eastward across the northern plains to the Midwest was drier. In fact, it seems likely that what is now the great American granary was then a dry grassland.

The evidence continues to mount, then, that American agriculture in particular is about to face a climatic challenge of unprecedented magnitude.

What *will* the headlines be in 2035? "GREAT PLAINS DROUGHT IN EIGHTH YEAR; GLOBAL FOOD CRISIS DEEPENS" . . . ?

Like a Fast-Moving Wave

There have been suggestions that farmers will adapt to climatic change, that—as our weather changes—they will move, plant different crops, modify varieties, and alter husbandry. There have been suggestions that the agriculture of some nations might benefit from a warmer world . . . that countries currently having vast areas of land too cold for large-scale farming, such as Canada and Russia, might come out as twenty-first-century winners.

All of this might be true if our climate were to change gradually, if we were to have time to plan, to cultivate different crops, to change methods, to seek government assistance. But our climate will not change gradually. Take a look again at figure 7.2 showing the extreme rapidity with which our world is forecast to warm. This is really a startling graphic, one you should keep in mind when people suggest that a greenhouse world might not be so bad. As James Hansen points out, "If we go all the way to a four-degree increase by the middle of next century, that will be an incredible climate change" occurring in an incredibly short time. Stephen Schneider puts it into quantitative terms when he says we're "altering the climate at ten to sixty times the natural rates." That is, over the next fifty years we're likely to warm our climate an amount that under natural conditions would have taken 500 to 3,000 years.

"We're altering the environment faster than we can predict the consequences," says Schneider. And Michael Oppenheimer, of the Environmental Defense Fund, metaphorically warns, "There will be no winners in this world of continuous change, only a globe full of losers. Today's beneficiaries of change will be tomorrow's victims as any advantages of the new climate roll past them like a fast-moving wave."

Even if farmers were able to react rapidly enough to keep up with the "fast-moving wave," that doesn't mean ecosystems could. Canada, for instance, probably couldn't take full agricultural advantage of a warmer climate, since much of the nation does not have the optimum-type soil for growing wheat and corn.

Despite the availability of modern technology and government support, one has to be skeptical regarding just how adaptable even the American farmer might be. Remember that in the 1980s, U.S. agricultural and livestock losses from drought and heat totaled over forty-five billion dollars. And that was just "spring training."

These losses have to be considered against a background of just how important farming is to the U.S. economy. Besides feeding us, U.S. crops bring in about thirty-nine billion dollars of foreign exchange annually. But the crops are not critical only to the United States. The Midwest has been described as the "world's safety net against starvation." In 1991, for instance, American farmers grew 41 percent of the world's corn and provided 29 percent of all the wheat and 20 percent of all the rice traded on world markets.

So it comes to this: a threat to American agriculture is far more than a threat to the U.S. economy; it is a threat to our ability to prevent large segments of the world's population from starving.

In the Sahel, which extends across the African continent from Senegal to Ethiopia, ten million people were forced from their homes by drought to search out food, water, and shelter during the 1980s. Ten million people. Tell them the greenhouse effect won't be so bad. Or tell it to the eighteen million southern Africans threatened by drought-induced starvation as recently as 1992.

What will the headlines be in 2035?

Other Problems . . . Water and Wildlife

Water: Demand vs. Supply

Chapter 4 outlined some of the water-supply problems we will face in a greenhouse world and focused on the near-term threat of water shortages in the Colorado Basin. But the Colorado isn't the only river basin where demand is likely to far exceed supply. The Rio Grande Basin, covering most of New Mexico and west Texas south of New Mexico, will be equally vulnerable—if not more so—to acute water shortages as temperatures soar and precipitation dwindles.

For that matter, much of the western United States will likely be in a mad scramble for every last drop of water as the greenhouse effect intensifies. The Great Basin (most of Nevada and western Utah), California, and much of Texas are all highly susceptible to demand outstripping river basin supplies.

The Missouri River Basin, which covers a broad area of the northern Great Plains, will also be faced with water shortages, although shortfalls there might not be so extreme as in the Colorado and Rio Grande basins.

Electricity Demand and Water Supplies

With a warmer world will come an increased demand for cooling, and thus for electricity to run air conditioners. For instance, an Environmental Protection Agency (EPA) report projects that in the United States, air conditioning by 2055 will suck up 14 to 23 percent more electricity during peak demand than would be the case without a climate change. This means an even greater number of generating plants than currently planned for would have to be built. That would certainly bring the battle between fossil fuel and nuclear power plant advocates to a head,* but it has significant ramifications relative to water supplies, too: vast quantities of water are required to cool power plants and process fuel.

Not only that, but some parts of the country are highly dependent

---

*The EPA report warns that if the unplanned-for generation capacity is met by fossil-fuel-fired utilities, an additional 250 to 500 million tons of CO_2 will be added to the atmosphere each year.

upon hydroelectric power. And the greater a region's reliance upon water power, the more vulnerable its electric supplies are to prolonged drought and dwindling river flows. In the Southeast, almost 30 percent of the electricity in the Tennessee River Basin comes from hydroelectric dams. Similar amounts are generated by hydro facilities in California and the Lower Colorado Basin. In Alaska, half the electricity comes from water power. But it is the Pacific Northwest that could be in the most precarious position in a greenhouse climate, for in Oregon, Washington, and Idaho, fully 93 percent of the power is hydroelectric.

Too Little, Too Much

On a global scale, water-supply shortages will be a ticking time bomb for nations already squabbling over common resources . . . Egypt and Sudan, for example. Both countries draw water from the mighty Nile, and both want more, Egypt to support its exploding population, Sudan to combat the desertification stalking its Sahel.

In other parts of the world, water problems will take on a different character: too much. If the GFDL soil-moisture model is anywhere near correct, India, with the possibility of greatly increased summer rainfall, will be faced with the spectre not of drought but of massive flooding. Rivers like the Godavari, the Narmada, and the immense Ganges could frequently roil over their banks, driving millions of residents from their homes and flooding vast stretches of rich agricultural land. The misery wouldn't end in India, however, for the Ganges empties into the Bay of Bengal through where? Bangladesh.

As Michael Oppenheimer says, "There will be no winners. . . ."

And already warning precursors for Bangladesh and India have struck. In September 1988, Bangladesh was swept by the worst monsoon floods in the country's history. Unofficial estimates set the death toll at 2,000; twenty-five million people were left homeless, and five million acres (over two million hectares) of rice land were inundated. A month later, record monsoon rains thundered down on northwest India, triggering heavy flooding. Casualties mounted into the hundreds; tens of thousands of farm animals drowned; and vast quantities of food stores were lost.

Contributing to the flooding potential on the Indian subcontinent is the deforestation of the Himalayas and their foothills. When the monsoon rains arrive, the stripped mountains along the roof of India and Nepal are subject to enhanced erosion that allows great surges of water to rush southward through the rivers of Bangladesh and northern India. So it isn't the devastation of only tropical forests that threatens our environment; the crisis extends into our midlatitude woodlands, too.

No Place to Go

Even flora and fauna will suffer in a greenhouse world. Wilderness areas are increasingly hemmed in by commercial development, and as the climate changes, fragile ecosystems will be unable to respond: plants (through seed propagation) and animals will have no place to go.

In the Arctic, caribou will eventually lose their migratory routes as ice bridges between islands melt. Farther south, as the sea level rises, marshes will attempt to shift slowly inland; but they, too, will be blocked by development: condominiums, levees, and roads. As the marshes, swamps, and bayous disappear, so will their abundant and diverse wildlife, and so will estuarine nurseries for fish, shrimp, crabs, and oysters.

In the tropics, mangrove forests, whose dense networks of roots and runners trap sediment and build protective islands, will also try to build inland as the oceans come up. But in places where development has reached the shore, the mangroves, like the marshes, will have no room to move . . . no room to survive.

On dry land, too, there is a limit to the distance a forest can relocate within a year. Michael McElroy, a planetary scientist at Harvard University, points out, "If [a forest] is unable to propagate fast enough, then either we have to come in and plant trees, or else we'll see total devastation and the collapse of the ecosystem." Human intervention might not help, though. While a suitable climatic area to which a forest could be relocated *might* be found, there's no guarantee the soils there could support the particular species of trees within the forest. And, of course, such relocation doesn't begin to address the wildlife problem.

Perhaps we will have this all figured out by the time the landscape in South Carolina begins to resemble that of South Dakota.

PART III

Regional Consequences: The United States

9/ The West

This chapter and the two following focus on specific greenhouse-effect consequences likely to occur in various regions of the United States over the next sixty years or so. This time frame has been chosen because it represents a period through which many younger readers of this book could live to see momentous climatic changes. Certainly my grandson, Nicholas, to whom this book is dedicated, can expect to be around through the middle of next century. I, in my middle age, might make it through the first quarter of the upcoming century. Even the more elderly perusers of these chapters may well experience the first incontrovertible manifestations of the greenhouse effect before the turn of the century.

The scope of the discussions relating to regional climatic consequences is necessarily limited. The more important implications for each U.S. area are addressed, but certainly not in the detail to which the Environmental Protection Agency (EPA) went in its three-volume, 750-page study "The Potential Effects of Global Climate Change on the United States."

Absent, too, are in-depth discussions relating to the effects of rising sea and Great Lakes levels. These effects have been examined by researchers, but primarily in response to climatic warming resulting from doubled atmospheric CO_2 concentrations. Since sea and lake levels respond relatively slowly to climatic warming, the headline-making disaster scenarios tied to rising seas are not likely to plague us for another century. This doesn't mean such eventualities needn't be of concern to us now, just that they don't fall within the scope of what I have chosen to write about: greenhouse-effect consequences we're likely to experience within our lifetimes. Where appropriate, I do talk about the ramifications of a two- or three-foot (1-m or less) sea-level rise by midcentury (see chapter 8).

The sources of information and data used for this chapter and the two that follow include the EPA report, the published research of Hansen and others at NASA and of Manabe and Wetherald at the Geophysical Fluid Dynamics Laboratory (GFDL), and an update of the work I did for *The Greenhouse Effect* using the climate of the 1930s as a near-term analog for the United States.

Regional Forecast Problems

While climate models handle expected climatic changes quite well on a global scale, they weren't really designed to support detailed studies of regional changes. This is because the "boxes" that climate models work with are quite large . . . remember, the NASA computerized model generates calculations for boxes only slightly smaller than the states of Utah, Colorado, Arizona, and New Mexico combined (chapter 7).

This means that any change calculated for the single grid point that represents a particular box must be applied to all points within that box, even though climatologists know that isn't exactly how the real-world climate will respond. For instance, when climatic changes for California were studied for the EPA report, changes for the entire state had to be defined based on *one* calculation point within California (and another one in southern Oregon). This was true for each of the three climate models—NASA, GFDL, and Oregon State University (OSU)—the EPA used in its analysis.

Despite this drawback, this approach is better than anything else researchers have, and forecasts carrying a relatively high degree of confidence can be made, especially when several different models come up with similar predictions for the same area.

This, in fact, is often the case for temperature predictions. Precipitation predictions, on the other hand, are frequently in conflict. This is because the complex physics of precipitation are more difficult to model than those of temperature variation, and because the precise ways in which oceans may influence changing precipitation patterns are not yet well grasped (chapter 7).

Improvements in climate model precipitation forecasting will occur as the role of oceans becomes better understood, model resolution improves (smaller "boxes" are used), and better representations of ground hydrology (how surface water is distributed and evaporated) and moist convection (vertical air motions in clouds) are made. Currently, different models handle the physics of precipitation with different degrees of detail.

The GFDL climate model is one of the most widely accepted when it comes to precipitation prediction (chapter 8), although the model's final output is not precipitation but soil moisture . . . which is really what farmers and water planners are most interested in. Soil moisture accounts for the combined influences of evaporation, snowmelt, and runoff.

Temperatures

Although several climate models predict that greenhouse warming will blanket all of North America, the models agree that less warming will occur over the western United States than elsewhere. That isn't exactly cause for celebration, though. For instance, the GFDL model suggests that by around the middle of the twenty-first century even the coastal and western valley sections of Washington, Oregon, and California will average 7° F (4° C) or more warmer during the summer than they do now. That would mean, to take one example, that summers in Portland, Oregon, would feel more like current summers in Sacramento, California.

Within about sixty years, the number of days with the mercury popping over 90° F (32° C) in downtown Los Angeles, California, may well have soared from five (the current average) to around twenty-seven, close to the number of such days that Denver, Colorado, now experiences. East of Los Angeles, in such spots as Ontario, Riverside, and San Bernardino, away from the moderating influence of the Pacific Ocean, the number of days topping 90° F would be considerably greater than twenty-seven, of course.

On a continental scale, too, the farther removed a region is from the ocean, the greater the projected warming. For instance, with doubled CO_2, the GFDL model predicts summertime readings over the Rockies to be roughly 11° F (6° C) higher than now. In more palpable terms, that means summers in Denver, Colorado, would be more like the modern-day scorchers that blister Dallas–Fort Worth, Texas. Denver now sees the mercury boil over 90° F about thirty times per year; by midcentury that number could almost triple.

Off to a Slow Start

Supporting the notion of less greenhouse warming in the West than elsewhere in the United States is the analog of the 1930s. It suggests the 1990s will bring little, if any, overall warming to the western third of the country (see figure 5.3). As a matter of fact, it hints that much of California, Utah, and New Mexico, as well as large parts of Nevada and Colorado, will average out slightly cooler. During the 1930s, wintertime temperatures in the intermountain region of the West averaged lower than current readings (summers, however, were generally hotter and drier). In California the lower temperatures were mainly a summertime phenomenon, especially noticeable near the coast (see figure 2.9).

In rough agreement with the analog, the NASA model under scenario A ("business as usual"—see chapter 7) foresees little change in winter temperatures over most of the western United States during the 1990s. It also predicts warmer summers (by 2° or 3° F [1° or 2° C]) across the Pacific Northwest eastward through Montana and Wyoming, little change in summertime readings along the southern California coast, and slightly cooler—or not quite so hot—summers in the Southwest.

After the 1990s, however, all of the West, along with the rest of the country, will likely grow steadily warmer. Even under NASA's more moderate scenario B ("limited emissions"), virtually all of the United States and Canada would average 2° to 3° F (1° to 2° C) warmer during the second decade of the coming century. The one exception to this would be in the Southwest (most of Arizona, Utah, and New Mexico), where the warming would be limited to 1° to 2° F (0.5° to 1° C).

Precipitation

Summer soil moisture, according to the GFDL model, will diminish across most of the United States through the early part of the twenty-first century. By midcentury, western soils in the summer are expected to be roughly 30 to 45 percent drier than they would be in a world with half as much atmospheric carbon dioxide. For California, another climate model suggests summertime precipitation might actually increase somewhat. But the markedly higher temperatures expected would lead to significantly greater evaporation than now occurs and thus to drier soils in spite of more rainfall.

All three models used in the EPA study indicate winter precipitation in California will be a bit greater by midcentury. If the results are averaged, the implication is that about an inch (25 mm) more of precipitation will fall from December through February. The GFDL model also suggests more winter rainfall in the Pacific Northwest eastward to Montana and Wyoming. That's the good news.

Here's the bad news. For California (and probably the Pacific Northwest as well) the higher temperatures expected by midcentury imply that less of the winter precipitation will fall as snow and more as rain. That means greater amounts of the wintertime precipitation will run off—especially if the increased rainfall comes in bursts of heavier showers as postulated by NASA researchers (see chapter 7)—as opposed to being captured in mountain snowpacks. The warmer climatic conditions also suggest that what snowpacks do form will melt earlier and faster. That means less spring runoff to western rivers and diminished water supplies through summer when water is most needed.

The 1930s climatic analog hints that except for parts of the interior West—much of eastern Washington, Idaho, Wyoming, Utah, and Colorado—annual precipitation isn't likely to diminish significantly through the 1990s in the Far West (see figure 2.10). In fact, slight increases are implied for parts of the Southwest, southern California, and Nevada. It should be noted, however, that since precipitation is quite sparse in these areas to begin with, a 10 percent jump in precipitation doesn't amount to a great deal of moisture.

Consequences

For most of the West, water-supply crises will come to the forefront as the greenhouse effect takes over. The implications for the Colorado Basin have already been looked at (chapter 4). In California, the problems will be no less acute. Even without a climate change, California will have trouble meeting its growing demand for water: planners and managers project that by 2010 they won't be able to meet the states's needs with today's resource system. A climate change . . . warming . . . will only exacerbate the crisis.

"The Worst of All Possible Worlds"

The problem for California is as follows. With more winter rain (as opposed to snow), there will be more runoff and a greater threat of floods. Existing reservoirs will not have the capacity to store the runoff for later use and at the same time provide adequate flood protection. Much of the runoff will have to be released, resulting in lower water deliveries during the dry summer months when water is most needed. Irrigation water deliveries could dwindle by as much as 50 percent under the most adverse conditions. Peter Gleick of the Pacific Institute for Studies in Development, Environment, and Security sums it up by saying, "California will get the worst of all possible worlds—more flooding in the winter, less available water in the summer."

Californians already know what water shortages bring: besides severe use restrictions, skyrocketing prices. During the 1976–77 drought (see chapter 2), a temporary pipeline was laid across the San Francisco Bay to bring water to dried-up Marin County. Water was outrageously expensive, but residents were willing to foot the bill . . . at least for a while. Next time it won't be for "a while." The seventies drought also fostered a drop in hydroelectric production, with generation plunging by 50 percent in 1976 and 60 percent in 1977. In 1991, after five years of dry weather, water rationing again had become a way of life throughout much of the state.

The threat to irrigation water supplies is perhaps the most critical. California produces over 10 percent of the cash farm receipts in the United States, mainly from cotton, apricots, grapes, almonds, tomatoes, and lettuce. In 1990 California farmers generated almost nineteen billion dollars in income. Crops in California are highly dependent on irrigation; as a matter of fact, farming accounts for 80 to 85 percent of the state's net water use!

In a greenhouse world with high CO_2, irrigated California crops—such as sugarbeets, cotton, and tomatoes—could do quite well, but only at the expense of demanding more water (due to the hotter weather) from an already stressed supply system. Nonirrigated crops such as corn could suffer in a hotter climate, with yield reductions exceeding 35 percent in some parts of the state.

Throughout the West, a significant percentage of cropland is irrigated. Thus it probably isn't unreasonable to assume that the problems California agriculture will face under a full-blown greenhouse effect will mirror the agricultural crises likely to confront many western states. All over the West soils will be drier in the summer, mountain snowpacks smaller in the winter.

There are many ways California *could* enhance its water supplies, but most are expensive and many have undesirable environmental consequences. New water-storage facilities could be built, but only at great cost and over the challenges of environmentalists seeking to protect wild and scenic river areas. Groundwater could be mined as a short-term solution, but in the long run this leads to problems such as land subsidence (as if California doesn't have enough to worry about with the San Andreas Fault). Southern California could attempt to obtain more water from the Colorado River, but as we have seen, the Colorado will have its own difficulties. Also, Arizona might have something to say about how Colorado River water is allocated. Desalinization plants could be constructed, but such facilities require a great deal of energy to operate and could contribute to air-quality problems and even the greenhouse effect. Cloud seeding is unproven and controversial, and could spark legal actions. In sum, none of the options to increase water supplies is attractive.

Elsewhere in the West there may be other, less-publicized ramifications of the greenhouse effect. For instance, in the Pacific Northwest, drier forests and higher fire danger (see the following section) could sharply curtail deer and elk hunting seasons; hunters may be restricted from entering tinder-dry woods. In Oregon, drier soils could bring an end to a twenty-million-dollar wild mushroom business. And of course with shrinking snowpacks because of warmer winter weather (higher freezing levels), the lucrative ski industry in the Cascades, Sierra Nevadas, and Rockies might be hard pressed to survive. It would seem that only the northernmost resorts in the highest elevations might have a chance of making it. (Perhaps the bulk of the business will merely migrate to British Columbia and Alberta.)

A Smoky Glimpse

A smoky glimpse of the future came to much of the Far West in 1988. Forest and range fires, fed by severe drought that left trees weak and underbrush and vegetation desiccated, consumed about 4.2 million acres (1.7 million hectares). Lightning fires ignited in late August gobbled up almost 800,000 acres (324,000 hectares) in northern California and over 100,000 acres (40,500 hectares) in southern Oregon. Smoke changed daylight to twilight throughout the region. And of course there was the great Yellowstone conflagration. All in all, the acreage incinerated in 1988 was more than triple that destroyed in 1987.

As the greenhouse effect takes over and summer soil moisture dwindles in the West, wildfire frequency will increase. Dry pine needles, discarded branches, rotted trees, and old brush will provide abundant fuel. The EPA goes so far as to speculate, "If dead wood rapidly builds up because of the decline in one or more tree species [due to climate change], large catastrophic fires could occur." In Washington and Oregon, the timber industry could suffer a significant decline.

Smoke from wildfires could become a common air pollutant and might even serve to briefly lower local air temperatures by blotting out sunlight. California, however, will have other air pollutants to worry about. As a result of greenhouse warming, the EPA study foresees a rise in August ozone concentrations in central California by midcentury.

Modeling results suggest a 30 percent increase in the number of August days exceeding the ozone air-quality standard. But in truth, that might be the least of California's worries by then.

Energy Demand

The greenhouse effect, with its sizzling summers, will bring an ever-increasing demand for electricity to cool homes and businesses. By 2055, on a national scale, the generating capacity required to meet peak demand—on the hottest days—could be 14 to 23 percent greater than currently foreseen (i.e., with no climate warming considered). The cumulative cost of providing this additional capacity could be as much as 325 billion dollars.

In the West, Arizona and New Mexico will likely be faced with the largest requirements for additional generating capacity to support air conditioning. In Arizona, demand by midcentury could be 20 to 30 percent higher than now expected; in New Mexico, 10 to 20 percent higher. These requirements for more generation capacity will bring with them additional baggage: even-greater demands for water. Unless we have evidenced the wisdom and foresight to develop solar energies by midcentury, new power plants will need water for cooling purposes . . . water that because of already overtaxed supply systems just might not be there.

In several of the more northern states of the West—Washington, Oregon, Montana, and Wyoming—decreased demand for heating because of milder winters will likely exceed the requirement for more summertime cooling. Thus in those states demand for additional capacity might actually be less than currently foreseen. That's certainly welcome news for Washington and Oregon, which are highly dependent on hydropower and faced with dwindling river flows in a warming climate.

Elsewhere in the West, midcentury demand for greater generating capacity will likely be moderately greater than now expected, generally less than 10 percent.

"There Must Be a Pony in Here Someplace"

Amid the dust from dried-out soils and smoke from western wildfires, some silver linings may be found. Some benefits might in truth be realized from the greenhouse effect. Perhaps, though, they would fall less in the silver-lining category and more under the heading of "there must be a pony in here someplace" (from the old story of an optimist and a pessimist locked in a room full of horse manure).

Assuming enough water can be foraged, western irrigated agriculture could certainly benefit from the inadvertent fertilization provided by more CO_2 and from the longer growing season resulting from climatic warming. Be mindful, though, that such benefit would come only at great cost. However—and if—the water supply challenge is met, it will be terribly expensive. The costs, naturally, will be passed along to us, the consumers, from the farmers. Food prices will soar and contribute to what could be a disastrous economic scenario in the United States. I talk about such a scenario in chapter 12. It's a scenario that can be avoided (but probably won't be).

A benefit wielding less of a double-edged sword could result from increased summer tourism and recreation, especially along the coasts of Washington, Oregon, and northern California. These coasts, because they are now relatively cool in the summer, have not seen the rapid development of recreational facilities that many other U.S. shores have. You have to be a hardy soul indeed to take a dip in the ocean off the Pacific Northwest. But with a projected summer warming of 7° or 8° F (4° C or more) by midcentury, the climate of Seaside, Oregon, may become more like the present climate of Malibu, California . . . with better surf.

All things considered, climatic warming might bring more than tourists to the Pacific Northwest; it could trigger a significant population migration to this far corner of the United States. As the climate of the Northwest transforms into one more Mediterraneanlike, the states of Washington and Oregon, with their abundant and varied natural resources and vast living space, could become the California of the twenty-first century. Oregon for many years, and ultimately to its own economic detriment, did not encourage such migration. In the future, the state may welcome growth—undoubtedly controlling it—to become party to a major economic power grab, along with Washington state, from America's present Sun Belt. Perhaps it won't be an earthquake after all that triggers the demise of California.

Uncle John

The trend toward warmer Pacific Northwest summers may already have begun . . . in a subtle way. My Uncle John, who lived for many years in Portland, Oregon, may have been one of the first to notice. When he reached retirement age he began to covet a move to southern California and warmer, drier weather. But he seemed in no hurry to leave Oregon. Once, in the early or mid-1980s, I asked him why, and he said, "The summers have been so nice here recently, I hate to leave."

As usual, meteorological curiosity got the best of me, and I did some checking. Sure enough, two-thirds of the summer months (June, July, and August) in Portland from 1979 through 1988 had temperatures that were above the long-term average. July 1985, with a mean of 74.1° F (23.4° C), was the warmest month ever (at the airport, where official records are now kept).

Uncle John and his wife, Dorothy, finally did move to California . . . to Palm Desert, where July maxima by midcentury may average close to 115° F (46° C). John and Dorothy will be long gone by then, of course. To someplace not nearly so hot, I trust.

10 / The Great Plains and Midwest

Nowhere is the greenhouse effect in the United States likely to be felt earlier or more severely than on the Great Plains and in the Midwest—America's breadbasket. It was here that the legendary Dust Bowl of the 1930s (see chapter 2) caused wheat and corn yields to plummet by up to 50 percent. It was here where, despite modern technology, the heat and drought of 1988 triggered a 40 percent reduction in midwestern corn yields (chapter 4). And it is here where the first major battle of our new greenhouse world will likely be joined: the Great Drought of the 1990s (the subject of chapter 2).

Temperatures

Based on the climatic analog of the 1930s, the northern and central Plains and western portions of the Midwest may well experience the largest annual greenhouse warming of any place in the United States during the latter part of the 1990s (see figure 5.3). The analog suggests yearly means will climb as much as 3° F (1.7° C). July is likely to manifest the greatest hike in temperatures, especially from Kansas City northward along the Missouri Valley into North Dakota. Those regions could bake under July readings averaging 4° to 5° F (around 2.5° C) higher than current means (see figure 2.9). In the hottest summers, monthly temperature departures could reach a withering 12° F (6.5° C) or more (see figure 2.1).

The NASA climate model, under scenario A ("business as usual"—see chapter 7), also depicts hotter summers during the late 1990s for much of middle America. The model suggests that summer (the combined months of June, July, and August) will average 2° to 3° F or more (1° to 2° C) hotter than recently over a broad area of the country centered on southeastern Missouri. Even under the more moderate scenario B ("limited emissions"), the NASA model predicts that by the 2010s, all of the Midwest and Great Plains will be averaging 2 to 3° F or more warmer *on an annual basis.*

As the greenhouse effect figures more and more prominently in our climate, each successive decade will grow warmer. By midcentury the

GFDL model foresees summers averaging an astounding 14° to 16° F (8° to 9° C) hotter over the northern Great Plains, northern Great Lakes, and southern plains of Canada. Summers in Fargo, North Dakota, would feel like summers do now in Dallas–Fort Worth, Texas.

Hot Summers and Corn

James Hansen and his co-workers at NASA came up with a good way of describing what the trend toward more torrid summers will mean in terms of the probability of any *one* summer being "hot." For Omaha, Nebraska, they arbitrarily defined the ten warmest summers (June, July, and August) in the period 1950 through 1979 as "hot," the ten coolest as "cold," and the middle ten as "normal." In other words, for that period there was a 33 percent (one out of three) chance of any one summer being "hot." Under scenario A the probability of a "hot" summer for Omaha rises sharply to 80 percent (eight out of ten) for the late 1990s, drops back a bit in the first decade of the new century, rises to 85 percent for the 2010s and 2020s, then locks in at 100 percent for the remainder of the century!

Things aren't much better with scenario B. The probability of a "hot" summer under that scenario is over 50 percent after 1995. From there it increases each decade, reaching 85 percent for the 2020s, the same as under scenario A. (No scenario B calculations were made for decades beyond the 2020s.) Even given the unlikely, relatively benign scenario C ("Draconian emission cuts"), the chances of a "hot" Omaha summer exceed 50 percent each decade after the turn of the century.

In terms of what farmers are more concerned about, consider the likelihood—as NASA researchers did—of five consecutive days with maximum temperatures over 95° F (35° C) in Omaha. Such conditions are thought to represent a threshold above which the productivity of corn suffers. During the period 1950 through 1979, at least one five-day heat wave with maxima exceeding 95° F occurred in three out of every ten years. Under scenario A this becomes five out of every ten years by the late 1990s and seven out of every ten years in the 2020s. Even if scenario A doesn't hold up into early next century, the outlook isn't a whole lot better given scenario B: by the 2020s six out of every ten years would suffer corn-killing hot spells.

The NASA calculations warn that for doubled CO_2, nine years per decade will experience such runs of fiery temperatures. In other words, near the middle of next century or shortly thereafter, virtually every summer around Omaha would be blistered by one or more severe heat waves.

Greenhouse Summers

By midcentury, give or take a decade, all parts of the Plains and Midwest most likely will have undergone a significant climatic change. The sweaty grip of greenhouse summers should be unmistakable. Chicago, Illinois, may experience the frequency of days over 90° F (32° C) that parts of the lower Mississippi Valley currently do. (The Chicago figure is expected to soar from sixteen to fifty-six such days per year.) In Dallas, Texas, the number of days over 90° F could approach that which Phoenix, Arizona, now has (the Dallas figure is forecast to escalate from 100 to 162 per year). And in Dallas it wouldn't be a dry heat.

On the meltdown side of the thermometer scale, consider the frequency of days over 100° F (38° C) in a world with doubled CO_2. In Omaha the number of days per year exceeding 100° F is predicted to be seven times greater than now (twenty-one versus three). In Chicago, which rarely suffers readings over 100° F currently, such sweltering weather is forecast to come along about half-a-dozen times each year by midcentury. In Dallas, where reaching the century mark is not uncommon even today, it will be even more common with twice as much CO_2 in the air (the annual tally of 100° F–plus days is expected to soar from nineteen to seventy-eight).

Greenhouse Winters

Don't get the idea, however, that even though summer heat will likely be more widespread and oppressive than ever by midcentury, cold weather will have disappeared. Certainly cold waves will be fewer and further between, and what frigid weather does occur will be less intense; but freezes and frosts won't be a thing of the past. In Omaha, for instance, the number of days with mimina below 32° F (0° C) is predicted to drop from 139 days annually to 75. In Chicago, the yearly frequency is forecast to plunge from 132 to 71 days; in Dallas, from 40 to just 11. Keep in mind, too, that in the near term, the 1930s analog suggests that frigid winters will still plague us through the 1990s. Remember, the coldest month ever in the United States, February 1936 (see chapter 5), directly preceded the hottest summer on record. (The lesson here is that a cold winter does not mean the greenhouse threat has lessened, any more than a single hot summer—such as 1988—is incontrovertible evidence that the greenhouse effect has taken complete control of our weather.)

Precipitation

The thirties analog implies that during the latter half of the 1990s, the state likely to be hardest hit by a *combination* of reduced precipitation and higher temperatures is Kansas (see figures 2.10 and 5.3), the leader in U.S. wheat production. Also likely to suffer a blend of significant drying and warming are large parts of Nebraska, Iowa, and the Dakotas. Other areas may endure it to a lesser extent: portions of Oklahoma, Missouri, and Wisconsin. Any amalgam of diminished rainfall and enhanced heat leads to drier soils, of course, and it is dry soils that most worry the farmers of America's heartland.

If any region of the Plains and Midwest is likely to escape the initial onslaught of the greenhouse effect, it is probably Texas. There the thirties analog suggests slightly greater precipitation through the 1990s (figure 2.10). Even that is likely to change by midcentury, though. The GFDL model indicates that within another sixty years or so all of middle America—with the possible exception of the Texas coast—will find drier soils a recurrent feature, at least in summer.

Extreme summertime drying—soil moisture reductions exceeding 50 percent—is suggested by the GFDL model for the far northern Plains and northwestern Great Lakes. Soils at least 30 percent drier are foreseen for the remainder of the northern Plains, across the central Plains, and throughout the Midwest.

Different Answers, Same Results

The three models examined by the EPA for its report come up with different answers regarding the magnitude and seasonal patterns of midcentury precipitation changes likely for the Great Plains. When considered on an annual basis, however, there is no disagreement relative to end results: all three models point toward diminished soil moisture. Two of the models foresee less annual rainfall and snowfall on the Plains, while one predicts a tiny increase. But given the sharp warming likely, such a slight increase would be quickly gobbled up by enhanced evaporation. Result: drier soils.

Manabe and Weatherald of the GFDL point out that the significant loss of summer soil moisture expected on the Great Plains would result from three major factors: (1) earlier melting of winter snows, (2) greater evaporation, and (3) less summertime rainfall. (The GFDL model isn't alone in forecasting dwindling summer precipitation on the Plains. The NASA model does, too; the OSU model, however, foresees a small increase . . . about a third of an inch [9 mm] over the entire summer. But even if that proved correct, it wouldn't be enough to stop the summer landscape from turning to dust.)

The three EPA-considered models also give different answers regarding midcentury summertime precipitation around the Great Lakes. But if the results of all three models are averaged, the net outcome is "no change." Of course, even under that scenario the blistering summer heat would win out, dictating drier soils.

On the good news side of things, the models all agree on one thing: on an annual basis, precipitation totals around the Great Lakes may well increase in response to doubled atmospheric CO_2. The GFDL calculations suggest that most of this increase will come in the winter and not be limited to just the Great Lakes. It will probably encompass much of North America from the northern Plains and Great Lakes northward through Canada. Thus winter soil moisture would be enhanced over much of this region, with the largest increases likely west of Hudson Bay across the Canadian tiaga and tundra. But that wouldn't help U.S. farmers.

Consequences

Across the Great Plains and in the Midwest, agriculture will suffer the most in a full greenhouse environment. (That's an ironic way to state it, I suppose.) The bad news for the latter part of the 1990s has already been delivered (in chapter 4): tens of billions of dollars in crop and livestock losses due to heat and drought. Don't expect it to get any better after that.

For example, although the GFDL-modeled precipitation decreases for midcentury in Kansas and Nebraska are no greater than those of the worst Dust Bowl years (1934 and 1936), temperatures are forecast to be substantially higher than during the Dust Bowl. And higher temperatures appear to be the primary culprits in expected crop yield declines. High temperatures shorten the period during which crops can mature.

One researcher, using a crop-yield model tuned to 1975 technology, determined that a recurrence of 1934 and 1936 climate conditions would more than halve wheat yields on the Great Plains. And this result did not even consider the hotter weather in the offing. Other researchers, looking at corn and soybean production in the Midwest and Great Lakes states, suggest that dryland (nonirrigated) crop yields could tumble by as much as 60 to 65 percent under the GFDL climate scenario for doubled CO_2. But under some of the wetter scenarios as foreseen by different climate models, crop yields in some northern areas—around Duluth, Minnesota, for example—could actually increase.

Mitigation?

Admittedly, a couple of things could help mitigate the extensive crop losses feared for the Great Plains and Midwest. One is CO_2 itself. An increase in carbon dioxide should enhance the growth of crops, but as discussed in chapter 4, the jury may still be out on that. (Researchers found that although soybeans grew more prolifically in a high-CO_2 environment, pests ate more of their leaves because the leaves were less nutritious.) The other mitigating factor is irrigation.

First, carbon dioxide. Although atmospheric CO_2 may well provide a fertilizing effect to crop growth, the EPA cautions that "experimental results from controlled environments may show more positive effects of CO_2 than will actually occur in variable, windy, and pest-infected (e.g., weeds, insects, and diseases) field conditions." The EPA also warns: "The more severe the climate change scenario, the less compensation provided by direct effects of CO_2." In other words, there is a threshold beyond which carbon dioxide won't help. If it's too hot and too dry for crops to grow, crops won't grow . . . no matter how rich in carbon dioxide the atmosphere is.

Now, irrigation. In many parts of the Great Plains, irrigated farming of corn, rice, and cotton has replaced dryland wheat production, especially in western Kansas and the Texas Panhandle. Overall, around 12 percent of Great Plains cropland is irrigated. In Nebraska, Kansas, and Oklahoma, about three-quarters of the water for irrigation is supplied by groundwater, most of it coming from a vast underground reservoir known as the Ogallala Aquifer. This aquifer, extending from Nebraska through the Texas panhandle, supplies irrigation to approximately fourteen million acres (5.7 million hectares) of land in Nebraska, Kansas, Oklahoma, Texas, Colorado, and New Mexico.

But forty years of heavy use has taken its toll on the aquifer, and it is being depleted of its water far faster than it can be replenished. Part of the reason for the rapid depletion is the switch by farmers to corn to satisfy the nation's appetite for corn-fed beef. Corn requires a 50 percent supplement of water relative to normal rainfall, while wheat and sorghum need only a 10 to 20 percent augmentation. In some parts of Kansas, Oklahoma, and Texas, the Ogallala is dropping three feet (one meter) per year. It is recharged in those areas at the rate of one-half inch (slightly more than a centimeter) per year.

Conservation efforts, along with hard times that have forced some farmers off their lands, have cut the drawdown rate of the Ogallala in half over the past decade. Still, the long-term outlook, even without a greenhouse effect, is sobering. Donald Reddell of Texas A & M University explains that although the aquifer will never be pumped dry, "what is going to happen is that the water yield, in gallons per minute, is going to drop, and because of energy costs it will reach a point where the farmer cannot afford to pump water to irrigate his crops . . . and [he] will stop farming."

Economic Impacts

The EPA, for its report, studied the economic consequences likely to result from changes in crop yields and water availability fostered by the greenhouse effect. The conclusion: "The results of the . . . study imply that wheat and corn production may shift away from the southern Great Plains, causing dislocations of rural populations. For many rural communities in the region, this may further weaken an economic base already under pressure from long-term structural changes in U.S. agriculture." The report goes on to warn of potential risks to financial institutions supporting farmers and to emergency relief resources.

In the Midwest and around the Great Lakes, because of the differing climate change scenarios foreseen by the models, it is difficult to predict whether crop yields would decline or rise. If the GFDL model results are correct, yields will fall and demand for irrigation water will burgeon. As a matter of fact, under the hot, dry conditions forecast by the GFDL model, corn production in Illinois—the second-ranking corn-growing state in the country—might become unfeasible. The EPA report speculates that farmers might react by installing irrigation systems (if they can afford them); planting short-season corn, soybeans, and sorghum; and probably abandoning marginal croplands.

Livestock

Not only crops but also livestock will suffer in a greenhouse world. Texas, for instance, raises nearly 40 percent of the beef cattle in the United States. Heat stress for these and other farm animals will increase markedly as the greenhouse effect becomes more pronounced. Recall that for Dallas, Texas, the number of days exceeding 100° F (38° C) is expected to quadruple (from nineteen to seventy-eight per year) with double the amount of CO_2 in the air. Livestock reproduction dwindles with increasing heat, and hotter weather may allow tropical diseases to extend their ranges into the Great Plains.

Livestock problems of a different sort might plague the northern Plains. Dried-up duck-nesting ponds could cause the populations of mallards and canvasbacks to plummet. Of course, indigenous wildlife in general will suffer in a hotter and drier climate.

Forests

Around the Great Lakes the distribution and abundance of forests will change as the climate grows steadily warmer. The range of spruce trees will gradually shift northward out of the region, while in Minnesota—within thirty to sixty years—the number of balsam firs will start to decline. Both of these tree species have commercial importance to the pulp and paper industry. Hardwood trees, such as oak, sugar maple, and yellow birch, will likely slowly replace the softwood evergreens as well as the white birch and quaking aspens.

Where oak, sugar maple, and yellow birch are relatively abundant now, mainly in southern parts of the region, the trend will be toward oak savannahs or perhaps even treeless scrubland sporting only prairie shrubs. Southern pines might eventually migrate into the region.

The Great Lakes

Although the levels of the Great Lakes will eventually fall substantially even in a wetter greenhouse climate, large drops in lake levels are not likely before midcentury. Ultimately, the level of Lake Superior may diminish by about a foot and a half (0.5 m), while the levels of Lakes Michigan and Huron could fall off by as much as eight feet (2.5 m) and that of Erie by as much as six feet (1.9 m). (No calculations were done for

Lake Ontario.) Since lake levels do not respond quickly to climatic changes, such huge drops are not likely to be attained until well into the twenty-second century, however.

By 2060, the level of Superior will probably have dipped only a few inches (several centimeters), while Lakes Michigan and Huron will likely be down about seventeen or eighteen inches (0.5 m). Even if precipitation totals increase with doubled atmospheric CO_2, lake levels would fall because higher temperatures bring smaller snowpacks—or cause them to disappear entirely—and greater evaporation.

A change more apparent than diminishing lake levels by midcentury may be a change in Great Lakes ice cover. Although ice on the lakes may not be totally eliminated, many winters could be virtually ice free . . . except on Lake Superior. On the other lakes, ice formation may continue to be common only close to shore and in shallow water.

Even though opportunities for ice fishing obviously would be limited, the news isn't all bad for fishermen. The warmer lakes generally would be beneficial to fish, since the higher temperatures would lead to more aquatic plant growth (meaning more food for fish) and a larger and longer optimum thermal environment. Translated, that means the temperatures fish like would be found to greater depths in the lakes, and they would be found during more months of the year. All of this points toward more and bigger fish.

Energy Demand

Throughout the Great Plains and Midwest, hotter summers will lead to ever-greater demands for cooling, and thus for electricity. By 2055, electric generating capacity to meet peak demand during the hottest weather could be as much as 30 percent greater than currently foreseen (i.e., no climate warming considered).

The largest jumps in peak demand would occur over the central and southern Plains (Nebraska, Kansas, Oklahoma, and Texas) and in Missouri. In those states, 20 to 30 percent more new capacity is likely to be needed by midcentury. In Illinois, a 10 to 20 percent increase in generating capacity—relative to what is currently expected—is probable.

In the more northern states of the region, winter heating requirements could dip sharply. But since the greatest summertime warming will likely be here, even these northern states would need additional electric generating facilities to meet peak demand (for cooling). Across the northern Plains and through the Great Lakes, up to 10 percent more new capacity would need to come on line by midcentury.

Michigan or Bust

On the Great Plains the requirement for new power plants may impose an additional burden on the already stressed Ogallala Aquifer. Electric utilities need water for cooling, and a potential source of that water on the Plains is obviously the Ogallala. This could bring agricultural and nonagricultural interests into direct conflict.

Or maybe everyone will just forget the whole thing and move to Michigan. The Great Lakes states, with milder winters and warmer summers, may well prove to have a more temperate climate than regions to the south come midcentury. And even though Great Lakes levels may well be dropping, the lakes will still contain vast amounts of freshwater. That in itself could be a big draw by the middle of the next century. "Consequently," the EPA study concludes, "the Great Lakes may be relatively more attractive than other regions." Like the Pacific Northwest, the Great Lakes states could become part of America's new Sun Belt in the twenty-first century.

Think of it. Spring break on the shores of Lake Michigan.

11 / The East and South

Among other things, we got tired of the long, cold New England winters after two decades. So in 1987 my wife and I moved from Boston, Massachusetts, to Atlanta, Georgia, in search of warmer weather. If only we could have hung on in Boston for another fifty years!

Temperatures

The GFDL climate model suggests that by midcentury or shortly thereafter, in response to doubled CO_2, summers around Boston will average 11° or 12° F (6° to 7° C) warmer than they currently do, or near 83° F (28° C). That would be a lot hotter than Atlanta summers are now (77.5° F or 25° C), and even a bit more torrid than current south Florida summers!

Of course, Atlanta summers will turn more oppressive, too; the GFDL model foresees an 8° F (4° to 5° C) increase in temperature. That means "Hotlanta" would really live up to its nickname, feeling more like Tucson, Arizona, with the steam turned on.

The greenhouse warming in the East may get off to a slow start, however. The 1930s analog hints that New England and much of the eastern seaboard down through Virginia could actually average out slightly cooler through the 1990s (see figure 5.3). But in the mid- and Deep South, annual temperatures could run 1° to 2° F (0.5° to 1° C) or more higher, mainly late in the decade.

Although 1990s summers could turn somewhat hotter over the South, especially from Georgia northwestward into western Kentucky, New England summers might average a tad cooler (see figure 2.9).

It is wintertime temperatures, though, that could display the largest warming after the mid-1990s; the thirties analog suggests that mean winter temperatures could be noticeably higher over much of the East and South. During the 1930s, Januarys in particular were much milder than currently, even across New England.

The NASA model, using scenario A ("business as usual"—chapter 7), also foresees greater winter than summer warming by the late 1990s for much of the East and South. For the combined months of December,

January, and February, the NASA model predicts at least a 2° F (1° C) warmup over the entire area, except for New England and New York. Warming exceeding 4° F (2° C) is suggested for Tennessee and the northern parts of Mississippi, Alabama, and Georgia.

Summers (June, July, August) after 1995, according to the NASA scenario A output, will average 1° to 2° F (0.5° to 1° C) hotter over most of the East, though New England might actually turn warmer than that. So too might much of the mid- and Deep South; the NASA implication is that most of Arkansas, Louisiana, Kentucky, Tennessee, Mississippi, Alabama, and a good part of Georgia will see summers hotter by up to 3° F or so (more than 1° C). Unlike the analog suggestions, the NASA model (scenario A) foresees no cooling over the eastern United States through the 1990s.

Even if we begin to cut back on greenhouse gas emissions in the near future, we still will have committed the eastern and southern United States—indeed, most of the nation—to an average annual warming of 2° to 4° F (1° to 2° C) by the 2010s (NASA scenario B—"limited emissions").

In response to doubled atmospheric CO_2, temperatures by around midcentury will be remarkably higher over the East and South. The OSU climate model—one of the most conservative—predicts mean annual temperatures will be up by 7° F (4° C) or more from northern Georgia and Alabama northward. As mentioned earlier, the GFDL model predicts midcentury summers will be 11° to 12° F (6° to 7° C) hotter around Boston (and most of the East), with slightly less warming in Dixie; e.g., 8° F (4° to 5° C) steamier around Atlanta, but less than 7° F (4° C) hotter over South Florida and in Louisiana.

Greenhouse Summers in the Nation's Capital

To depict what the trend toward more torrid summers will mean in terms of the probability of any single summer being "hot," Hansen and his colleagues did the same sort of calculations for Washington, D.C., that they did for Omaha, Nebraska (chapter 10). For Washington, as for Omaha, they arbitrarily defined the ten warmest summers from 1950 through 1979 as "hot," the ten coolest as "cold," and the middle ten as "normal." Thus for the period 1950 to 1979 there was a 33 percent (one out of three) chance of any one summer being "hot." Under scenario A the probability of a "hot" D.C. summer jumps to 70 percent (seven out of ten) for the late 1990s, reaches 80 percent in the first decade of the twenty-first century, and 100 percent by the 2020s . . . where it remains.

Under the more moderate scenario B, the chances of a "hot" summer in the nation's capital are in the 50 to 55 percent range for the latter part of the 1990s and the following decade. They reach 70 percent in the 2010s, then jump to over 80 percent for the 2020s (calculations stopped there). Even with the extreme greenhouse gas emission cuts imagined in scenario C (the "Draconian" one), the probability of a Washington sizzler exceeds 50 percent for the late 1990s, 2000s, and 2010s, then elevates to about 65 percent for the 2020s and 2030s.

The notice to be highlighted here is that *even were we to take drastic steps now to curtail the greenhouse effect, model results warn we've already committed the earth to a palpable climatic warming.*

Feeling the Heat

Given that we do nothing or very little to mitigate the greenhouse effect (NASA scenario A), New York City, in terms of days exceeding 90° F (32° C), will feel more like Washington, D.C., does now by the 2030s. (The number of days over 90° F is expected to double, going from about fifteen per year to thirty-three.) Washington, in turn, would suffer summers like the current ones that bake northern Mississippi (the tally of days above 90° F is forecast to soar from around thirty-five to sixty-seven annually). And in Atlanta, the number of days with highs topping 90° F would approach that which Jacksonville, Florida, now has. (The Atlanta tally is predicted to more than double, going from thirty-two to eighty-four per year.) No longer would Atlanta's elevation be a saving grace in the summer.

With twice the amount of CO_2 in the air, readings in excess of 90° F will be even more common by midcentury or soon after. In New York the tally could be up to forty-eight such days each year; in D.C., eighty-seven; and in Memphis, Tennessee, 145 (up from the current mean of sixty-five).

Maxima above 100° F (38° C) will also be more common with doubled CO_2. Here Memphis could really shine—or glow—with a tenfold increase in the frequency of such debilatating heat. (The yearly average is forecast to go from four to forty-two!) New York rarely sees the mercury top 100° F, but with twice the amount of carbon dioxide in the atmosphere it would happen on about four days each year. At the same time, Washington would bake in triple-digit heat a dozen times annually (the current average is but once per year).

While cold weather won't disappear entirely with doubled CO_2, the number of subfreezing days will dwindle substantially. By around midcentury in New York the frequency of days with readings below 32° F (0° C) will be about thirty-five annually (compared with the current mean of seventy-five). In Washington the yearly mean will be down to around twenty-eight (it is now seventy-two), while in Memphis it is predicted to be near fourteen (it is fifty-eight currently).

Precipitation

The 1930s analog suggests no significant drying over the East and South during the 1990s, except perhaps for Arkansas and northern Mississippi (see figure 2.10). Southern Florida, on the other hand, could well turn out to be wetter. Keep in mind, though, that the implication of no significant reduction in ten-year precipitation averages doesn't mean that shorter-term droughty periods can't occur. During the thirties, record dry years plagued parts of many states from Maine to Florida to Mississippi.

By midcentury the GFDL model suggests soils all over the East and South will be drier in the summer. Drying in the range of 10 to 35 percent is foreseen, with the greatest moisture loss over inland regions.

Varied Answers

For the southeastern United States the three climate models evaluated by the EPA in its study give varied answers regarding midcentury precipitation patterns. If the results of all three models are averaged, the implication is that on an annual basis, total precipitation won't change much. Even under this circumstance, however, soils would be drier because of higher temperatures and greater evaporation.

The biggest difference among the models occurs for summertime rainfall predictions. The GFDL model suggests that total precipitation for the months of June, July, and August will be about an inch and a quarter (32 mm) less than it is now. The other two models foresee an increase in rainfall. The NASA model, for instance, predicts that summertime showers and thunderstorms will lead to seasonal totals of more than two inches (over 50 mm) above current means.

Drier or wetter. Either scenario would bring summertime problems to the Southeast.

Consequences

Dry Weather Problems

Under the hotter, drier environment foreseen by the GFDL model, agriculture in the southeastern United States would be hit hard. Reductions in soybean yields—even considering the inadvertent fertilization provided by larger CO_2 concentrations—would range from 20 to 90 percent, with the biggest yield losses likely in parts of North Carolina and Georgia. Corn crop yields would dive by as much as 80 percent in some parts of the Southeast. And even irrigated corn would suffer yield reductions approaching 25 percent in certain areas, mainly due to higher temperatures.

Obviously, if the wetter NASA scenario held true, farmers would be much happier. Nonirrigated crop (soybean and corn) yields, encouraged by more CO_2 in the air, would increase in most regions, although some slight yield reductions likely would occur in parts of Alabama, Mississippi, and Tennessee. Crops currently irrigated wouldn't necessarily benefit from more rainfall; in fact, they might suffer a bit because of the broiling greenhouse heat.

If the hot, dry conditions predicated by the GFDL model develop over the Southeast, the EPA report notes that "a major increase in irrigation probably would be inevitable." Although water is currently plentiful in the region, the capital costs of irrigation are very high. This prevents most farmers from taking advantage of it; if and when they do, their expense surely would be passed along to consumers.

Dry weather, beyond reducing crop yields, also encourages the growth of certain weeds, both land-based and aquatic. For example, bitterweed is a tall, lacy-leafed, drought-tolerant plant that crowds out desirable pastureland grasses during periods of dry weather. When consumed by cows, it turns their milk bitter. The weed took over many fields in north Georgia during the 1988 drought.

Hydrilla and Eurasian water milfoil are aquatic weeds. They gain footholds in artificial lakes during times of low water. During recent years they have choked tens of thousands of acres in Tennessee Valley Authority (TVA) lakes. The weeds, which are virtually impossible to remove once established, eventually destroy game fish habitats and clog pipes drawing water for domestic and industrial use.

Hot, dry conditions and the decreased streamflows and lake inflows they imply would not only encourage weed growth but would also reduce hydroelectric generation, disrupt river navigation, and curtail or halt recreational use of many lakes such as heavily used Lanier northeast of Atlanta.

Wet Weather Problems

The wetter climatic scenarios predicted by some models for the Southeast would lead to problems of a different nature: flooding. The EPA report examined what could happen along the Tennessee Valley during a few exceptionally wet years around midcentury. The report notes that water storage would be inadequate at tributary reservoirs and that this could lead to uncontrolled spillage over TVA dams. Of greater concern would be "a high probability of flooding" at Chattanooga, Tennessee. The study warns that the two worst-case floods foreseen for Chattanooga could top river banks there, resulting in 200 million to one billion dollars in damages!

Other regions vulnerable to severe flooding problems would be southern Florida and central Louisiana. After severe rainstorms or river surges in those areas, water sometimes lingers for weeks on low-lying land.

Hot Weather Problems

As the greenhouse effect takes command of our climatic environment, the resultant mild winters and sizzling summers will in and of themselves bring problems, regardless of which direction precipitation totals go.

Some of the consequences relative to agriculture already have been discussed. But there are others, perhaps less obvious. For instance, while one might think the advent of milder winters would be universally welcomed, this wouldn't be the case with Georgia peach farmers. Peach trees need a certain number of hours below 45° F (7° C) each winter or they won't bear fruit. The application to the trees of a spray called elgetol can reduce the number of chill hours required, but such chemical use is not popular with consumers. The lack of chill hours became a concern to Georgia peach growers during the mild winter of 1988–89, a winter that was likely a precursor of many to follow.

A warmer climate will also encourage the northward spread of agricultural pests. Such insects as potato leafhoppers, sunflower moths, green cloverworms, and black cutworms will expand their ranges to the north by a couple of hundred miles (several hundred kilometers). If this colorful collection of critters that hop, fly, and crawl isn't familiar to you, fire ants might be.

Fire ants are tiny, mean-spirited insects that since the 1940s have spread across the South from Texas to North Carolina. Their stings are painful and in some cases fatal; about 21,000 people per year are treated by doctors for ant stings. The nasty little fire ants, natives of South America, are currently limited to regions where temperatures rarely fall below 10° F (-12° C). But as our climate inexorably warms, they will march northward through Arkansas, Tennessee, and the remainder of North Carolina. (They may already be doing this by crossbreeding and producing a hybrid ant that can withstand colder winter weather. Perhaps by midcentury the hybrid—responding to milder winters—will be at home near the Great Lakes and in New England.)

There is an exception to the generally bad news about agriculture and various pests. Agriculture in Florida might actually derive some benefits from a warmer climate and the diminished threat of freezes. Crops such as sugarcane—now largely found in the Everglades—and citrus could expand their ranges to cover the entire state. Certain other tropical crops, such as coffee, now just beginning to gain a foothold in Florida, could also be helped.

Overall, though, the outlook is not encouraging for agriculture in the Southeast. Among other things, marginal farmland will likely be abandoned, and this has some surprising implications for the landscape of the region. In the past, forests have been cleared for farms, and when farming ceased, the land reverted to forest. In the future, with a hotter climate, this probably will no longer be the case. Forests may find it extremely difficult or even impossible to regenerate from bare ground; abandoned land may become merely grassland or overgrown with weeds.

Southern forests in general will not fare well in a greenhouse climate. One forest growth model foresees a significant decline in forests from Mississippi to South Carolina between 2030 and 2060. Noncoastal areas, as around Vicksburg, Mississippi, and Atlanta, may suffer particularly large losses. Farther north, in slightly cooler regions such as Tennessee, the dieback may be smaller.

Given both the general decline of southern forests and the fact that they will not regenerate easily on abandoned acreage, the entire character of the landscape will likely be changing by midcentury. As the EPA suggests, "the [scenery of the] Southeast could gradually come to resemble the scenery found today in the Great Plains."

Was the naming of Plains, Georgia, prescient?

Hot Water Problems

With hot weather comes hot water: in rivers, lakes, and oceans. Coastal estuaries from Louisiana to North Carolina will grow warmer and warmer. At some point after midcentury a great proportion of their oysters, shrimp, and crabs—which cannot tolerate temperatures much higher than exist now—will have died or disappeared. One study predicts close to a 100 percent mortality for shellfish in Florida's Apalachicola Bay.

Shrimp and other mobile species might try to adapt by fleeing estuaries for cooler ocean waters in the summer, but in the open sea they would become fair game for predators. Other sea life likely to suffer significant mortality rates are spotted sea trout, oyster larvae, panfish, and flounder.

On the other hand, a few ocean creatures might benefit from the greenhouse effect. In a warmer Apalachicola estuary, for instance, pink shrimp, some finfish, and rock lobster might proliferate.

Even freshwater shellfish will be affected by a warmer climate. During recent droughts in the Southeast, large numbers of mussels in TVA lakes died. Record low water levels and searing heat robbed lake waters of dissolved oxygen without which the shellfish—worth an estimated sixteen million dollars—can't survive. The mussels are much sought after by the Japanese, who use them in their cultured pearl industry.

Energy Demand

Although requirements for winter heating will plunge as the greenhouse effect takes over, in only a few states—Maine, New Hampshire, and Vermont—will peak demand for electricity by 2055 likely require less generating capacity than currently foreseen (i.e., without consideration of the greenhouse effect). Elsewhere in the Northeast—with the exception of New Jersey—and in West Virginia, increased demand for summer cooling will likely drive capacity requirements up as much as 10 percent over those currently expected. In New Jersey, Delaware, Maryland,

Virginia, Kentucky, Tennessee, and Alabama, the increase could be even larger: 10 to 20 percent.

Throughout the remainder of the South, sweltering summer heat will send the demand for electricity to power air conditioning skyrocketing. By midcentury, new generating capacity needed to meet peak consumption during heat waves could be 20 to 30 percent above that now foreseen. In the Southeast alone, additional utility construction could cost 100 billion dollars more than currently expected.

Myrtle Beach, Maine?

In new England the trend toward a greenhouse climate will bring a gradual demise of the skiing industry. By around midcentury, only ski areas near the Canadian border are likely to survive, and then only with the aid of fabricated snow; incursions of warm rain will be more frequent and snowstorms fewer and further between.

Summer recreation in New England might benefit from climatic warming, however, as people flee the stifling heat of the Midwest and South. The Green Mountains of Vermont and New Hamphire's White Mountains might be particularly attractive. The coast of Maine, now cool and often foggy through the summer months, might also become a summer tourist mecca. By the middle of next century, summer temperatures from Kennebunk to Eastport could be more like those now found along the Carolina coast.

Farther south, winter tourism might pick up substantially along the beaches of Georgia and the Carolinas. People fleeing the "cool" winters of the northern states and Canada will likely discover they don't have to journey all the way to Florida or the Bahamas to find balminess.

Rising Oceans, Shrinking Beaches

As pointed out in chapter 8, the sea level is rising and the rise is going to accelerate. By midcentury, ocean levels are likely to be two or three feet (1 m or less) higher than they are now. Recall that in a greenhouse world, hurricanes along the Atlantic and Gulf coasts will have the distinct potential of being more severe than any we have known (chapter 3). So, considering the combination of elevated seas, more violent tropical cyclones, and ever-expanding coastal development, a conclusion that can't be ignored is that future hurricane damage will often be immense. But it won't take a two- or three-foot sea-level rise or a few blockbuster storms to alter the character of our beaches and wipe out oceanfront construction. An innocuous one-foot (0.3-m) rise in ocean waters will do that!

As ocean levels go up there is an initial loss of shoreline due to inundation. But the inundation doesn't stop there. A rise in sea level implies that the offshore ocean bottom must also rise (this is known as the "Brunn Rule"). To take care of this, ocean waves erode—from the upper part of the beach—the sand necessary to build the offshore bottom. The result is an even further reduction of beach width.

The amount of beach lost by erosion depends on the average slope of the entire beach from its upper reaches to its offshore bottom. Studies have shown that in response to a mere one-foot rise in sea level, beaches in the Northeast and along the shores of Maryland would erode back 50 to 100 feet (15 to 30 m); in the Carolinas, 200 feet (60 m) of beach would be lost; and in Florida losses would vary from 100 feet to as much as 1,000 feet (305 m). Over the flat wetlands of Louisiana the retreat of shorelines would be measured not in feet (meters) but in miles (kilometers).

Since most U.S. recreational beaches are less than 100 feet wide at high tide, it's obvious that even a one-foot increase in ocean levels would require human intervention to save beaches. And you can bet that won't be cheap.

Maine has partially addressed the problem through law. It has approved regulations allowing coastal development, but with the understanding that if seas rise and inundate property, the property reverts to nature. The cost of dismantling or moving flooded structures would be borne by property owners.

In the near future more states will follow Maine's example, for many greenhouse-effect challenges will have to be met on a regional basis. To be sure, those challenges will be immense. But on a national scale, they could be overwhelming.

PART IV

The American Challenge

12 / Converging Crises
Climate and Energy

The United States is setting itself up for at least two crises of immense proportions early next century. One crisis is, of course, the greenhouse effect. The other is a reprise of a crisis we've already been through, but from which we apparently learned nothing: the oil price explosions—and associated embargoes—set off by the Organization of Petroleum Exporting Countries in the 1970s. If you're old enough, you'll recall sitting in long lines at gas stations, hoping to get to the pumps before they ran dry. I recall searching frantically for antifreeze one autumn in New England, then paying an absolutely outrageous price when I finally did locate some. That seems a long time ago now, and it's hard to remember how it all came about. As philosopher George Santayana warned, "Those who cannot remember the past are condemned to repeat it." I believe the United States will soon verify that.

The ironic thing is that though it seems we will be threatened by two separate crises, both share the same roots: our use of energy. Theoretically, then, by addressing one crisis we address the other. What a marvelous opportunity. By buying our way out of one crisis—and it won't be cheap—we can buy our way out of two!

The Climate Crisis

The climate crisis, as has already been extensively discussed, will be manifest in the United States mainly through a series of agricultural disasters. It won't be an agricultural Armageddon, but remember, even in the 1990s, crop and livestock losses exceeding sixty billion dollars seem likely (chapter 4). From there it should get only worse. William Cline, a senior fellow at the Institute for International Economics, estimates that by midcentury, climate stress could cost American agriculture eighteen billion dollars per year!

Let me again summarize what may well lie in store for us. The Geophysical Fluid Dynamics Laboratory climate model suggests that by

2050 virtually all of the United States will suffer from depleted soil moisture in the summer, with most of the Great Plains and Midwest at least 30 percent drier than now. Dust and desiccation won't be the only scourges, either. The summer heat will likely be withering. The GFDL model predicts that June through August temperatures will average at least 7° F (4° C) higher across most of the nation, with a large area of the country from the northern Plains to the East Coast even hotter.

Joel Smith of the Environmental Protection Agency's Office of Policy Analysis, in a briefing to EPA officials, suggested that under such conditions the country probably would still have sufficient productive cropland to take care of domestic needs. But, he added, food prices would soar, and there would be few agricultural products available for export. That means we could kiss goodbye to almost thirty-nine billion dollars of foreign exchange income annually.

Foreign exchange doesn't hit home, however. What hits home is the size of the check or the amount of cash we hand over to the checkout clerk at the grocery store. Nine dollars for a big box of cereal? Eighteen? How about five bucks—or ten—for a pound of chopped beef? Maybe $4.50 for a loaf of bread, maybe $8.99. A pound of peaches? How's $3.30 sound? Or $6.50? Where might it all end? I don't know, but I'm afraid we're going to find out.

It will be costly and it will hurt, but we'll survive. I'm not sure about tens of millions in the African Sahel and Bangladesh. Or do we even think about that? If we are the "world's safety net against starvation," don't we have a humanitarian obligation beyond filling domestic tummies? In other words, shouldn't we begin viewing the greenhouse effect less in terms of national ramifications and more in terms of international challenges? Or will P.C. Snow be right: "The most dreadful thing of all is that millions of people in the poor countries are going to starve to death before our eyes . . . upon our television sets."

It's fine to look at an EPA report and say, "Well, I guess we can make it. We'll probably pay a price, but we'll still have food." Unfortunately, that doesn't address the global scope of the problem; it doesn't absolve us from dragging our feet toward energy innovation and independence . . . things that are demonstrably good for us anyhow! But we seem determined to be carried kicking and screaming into a world where fossil fuels must play a greatly diminished role.

Obviously the United States alone can't solve the problem. But we consume over a quarter of all the oil used in the world and thus have a moral duty to lead the way toward a fossil-free globe. Yes, it will cost a lot. It will cost a lot now; it will cost an order of magnitude more later, as time runs out.

Why not begin to take serious action before our destiny is manifest . . . if not to forestall the decline of American agriculture and preclude widespread starvation in the Third World, then at least to avoid the onrushing energy crisis?

The Energy Crisis

The energy crisis is a "stealth" crisis; it's creeping up on us unseen and unheard. Well, not exactly unheard; there have been a few Cassandras, but in the midst of plentiful oil, who worries about being held hostage to petroleum?

The facts are pretty straightforward. By the end of this century we will be importing a greater percentage of our oil than ever before (at least 50 percent, perhaps as much as 60 percent) and will be—either directly or indirectly—dependent on the powderkeg Middle East, where two-thirds of the world's known oil reserves lie, for a significant portion of our energy supplies. Within the first decade of next century—if not sooner— the abundance of oil will have expired, and guess who will again be calling the shots. Right . . . OPEC.

OPEC and Oil Prices—a Brief History

In mid-1992, the price of a forty-two-gallon barrel of oil on world markets was near $21. At the end of 1969, the price was in the $1 to $1.20 range, but the stage was already being set for massive increases.

Between 1967 and 1970, three events took place that formed the threshold for the great oil price shocks soon to ripple around the world: (1) as a result of the 1967 Arab-Israeli War, the Suez Canal was closed; (2) major oil companies underestimated demand for oil in Europe; and (3) a Syrian bulldozer severed the Trans-Arabian Pipeline, thus making Libyan crude, close to Europe, more valuable. The end result was that Libya was able to demand and get higher prices for its product. Using the Libyan success as a springboard, OPEC, in early 1971, claimed its first clear-cut victory over the West: a fifty-cent increase in the price of a barrel of oil. But this was merely a small taste of what was to come.

From 1971 to 1973, prices rose gradually. But the Middle East Yom Kippur War of October 1973 spurred Arab OPEC members to chop crude production by five million barrels daily as a rebuff to U.S. support of Israeli forces. At the same time, an enraged Saudi Arabia ordered an oil embargo against the United States and several other Western nations; most other Arab OPEC members followed suit. The result was a huge leap in the price of oil. In January 1974, OPEC's official price for a barrel was set at $7 compared with under $2 prior to the war. Spot market prices (for oil not covered by long-term contracts) went as high as $11.

The Arab oil embargo, which lasted six months, vividly demonstrated the vulnerability of the Western world to Middle Eastern events and OPEC control. OPEC's take per barrel of oil for the entire year of 1974 reached about $10, a 565 percent increase!

Between 1974 and 1978, oil prices eased upward, but less than the overall rate of inflation. Demand for OPEC crude dipped slightly because of slow economic growth, increased North Sea and Alaskan production, and conservation.

In late 1978, yet another event occurred that shook the oil-consuming world. Iranian oil workers shut off production in support of the revolution to overthrow the Shah; the resultant loss of five million barrels per day turned a small global surplus into a shortage. Although the Iranian cutoff was short-lived, OPEC took advantage of the situation to flex its newfound muscle and again raise prices. By early 1979, "marker crude" (Saudi Arabian light) was pegged at over $13 per barrel, but prices on the spot market soared to over $21.

Prices continued to escalate until 1981, when they reached as high as $35–$40 per barrel. But an economic slowdown (in part triggered by the OPEC price hike) and energy conservation (also induced by the OPEC price increases) led to a world glut of oil, and prices tumbled to as low as $12 per barrel in 1986.

Then, in the wake of Iraq's invasion of Kuwait on August 2, 1990, oil prices skyrocketed to over $40 a barrel, as Saddam Hussein's forces put themselves in position to control half the world's oil reserves (11 percent in Iraq, 11 percent in Kuwait, and 28 percent in Saudi Arabia, along whose border Saddam's forces were poised).* But after the successful

---

*Many Americans perhaps have little appreciation for how vulnerable the Saudi oil fields were during the first few weeks after the takeover of Kuwait by Iraqi forces. Saddam's army could have swept down the gulf coast to the Arab Emirates or overland to Riyadh, the Saudi capital, within a matter of days. Despite headlines touting a rapid buildup of U.S. troops, only token elements were arriving in August. One Bush Administration official said "we were scared to death" that Saddam would "figure out that he didn't have to hold the [Saudi] oil fields; he just had to blow them up. There was no way we could have stopped

prosecution of Desert Storm and the liberation of Kuwait, prices rapidly fell back to around $20 per barrel in early 1991.

U.S. Dependence on Imported Oil—Another Brief History

Oil currently supplies over 40 percent of all U.S. energy requirements. Just before the 1973 Yom Kippur War, the United States was importing about one-third of the oil it consumed. By 1979, at the time of the second big OPEC price boost, our dependency on imported oil had risen to 43 percent, with 70 percent of that coming from OPEC.

By 1983, however, as a result of energy conservation and slower economic growth induced by the steep OPEC price hikes, the United States was importing only 28 percent of its oil. Of that 28 percent, only two-fifths was from OPEC. Since then the import trend has dramatically reversed. By 1990 our dependence on foreign oil had risen to around 42 percent and was still climbing. Experts see the import figure reaching 50 percent by the mid-1990s and as much as 60 percent by late in the decade. This would be well above the level that existed in 1979. To be sure, we are less reliant on OPEC oil now than in 1979, but not by much. In 1990, 60 percent of our imports were from OPEC, and the figure was increasing.

A burdensome U.S. reliance on imported oil seems destined to continue. Our demand for energy, by conservative estimate, is likely to be almost 20 percent higher by the turn of the century than it is now. Consider this against the backdrop that—according to Jessica Tuchman Mathews of the World Resources Institute—90 percent of all the oil that will ever be produced in the lower forty-eight states will probably be gone by 2004. *All* domestic supplies, including ones from Alaska's North Slope, will have dried up by 2025. Although new discoveries are possible, they would add only a few years to our reserves.

On the other hand, OPEC nations possess 80 percent of the world's known petroleum reserves; 65 percent of the reserves are in the Middle East, and most of those are beneath the sands of Saudia Arabia. Thus unless the United States can somehow significantly reduce its depen-

---

him." Fortunately, Saddam had no satellites or spy planes to watch the U.S. buildup; he got much of his intelligence from Cable News Network (CNN). So U.S. military officials made sure television crews were shooting the arrival of every military transport landing in Dhahran, making it seem as if entire divisions were disembarking. The army's 24th Mechanized Infantry Division finally did arrive in early September, but some analysts feel it wasn't until November that allied nations had ground forces sufficient to deter a determined Iraqi thrust.

dence on imported oil, we will continue to be at the mercy of events either initiated by or occurring in OPEC nations, particularly those of the Middle East. Oil from the Persian Gulf now comprises over 13 percent of all our imports.

And ironically, the Gulf War, which should have come as a warning to us, probably served only to increase our reliance on OPEC. Both Kuwait and Saudi Arabia, in order to cover their war debts, upped their pumping after the conflict, an act likely to keep crude prices relatively low into mid-decade. Thus there is no monetary incentive for the United States to conserve energy or develop alternative, nonfossil sources.

The Threat

Melvin Conant, an energy consultant writing in the *Washington Post*, put the problem into blunt perspective, "As world oil demand inches up because of economic and population growth, and as economical reserves outside the Middle East become less certain, the world will become more dependent on a handful of unpredictable nations for its oil supply. That's not a forecast; it's a description of what is already taking place."

Robert Beckstead, professor of economics, Industrial College of the Armed Forces, notes, "There is no more serious threat to the long-term security of our Nation and allies than that which stems from growing deficiency of secure and assured energy resources." William Boseman of the Science Policy Research Division, Congressional Research Service, Library of Congress, looking ahead to the twenty-first century, predicts that "whichever nation or group of nations is dominant in energy at that time will also be dominant economically and militarily."

The United States is in anything but a dominant energy position now. Consider that from 1972 to 1980 our foreign oil tab soared from four billion to ninety billion dollars, an absolutely crippling increase . . . as we found out. By 1990 our imported oil bill had dipped to forty-two billion dollars, but that still was a substantial contributor to our 100-billion-dollar net trade deficit.

A number of scenarios can be developed that highlight the extreme vulnerability of U.S. oil supplies to external events. In the volatile Middle East, despite the military successes of the Gulf War, Saddam Hussein remained in power. In early 1992, Iran was buying Russian attack submarines, apparently with the aim of controlling the narrow Strait of Hormuz leading to the Persian Gulf. The Islamic republic was also taking delivery of a squadron of fighter aircraft (MiG-29s) from the former Soviet Union and importing Scud missiles from North Korea. And even in countries friendly to the United States, no significant moves toward more democratic regimes have come, leaving the door open to Islamic fundamentalists to foment unrest and revolution.

What Kuwait and Saudi Arabia both fear most is the export of the Iranian Islamic revolution. Saudi Arabia is dominated by the conservative Sunni Islamic sect, Iran by the radical Shia sect, which seeks revolutionary change. Doctrinal differences between the two sects have existed for centuries, and a fierce animosity between Shia and Sunni Moslems has developed. The Saudis point to the bloody clash between Saudi soldiers and Iranian pilgrims at Mecca during the 1987 hajj (pilgrimage) as an example of the lengths to which radical Iranian factions will go to advance dreams of Shia hegemony. Old fears in the area cannot be entirely laid to rest, either. A flareup of the simmering Israeli-Arab hostility could lead to a repeat of the 1973 Arab oil embargo against the West.

All in all, a more precarious dependence by the United States and its allies on remote politics and geography could hardly be imagined.

Oil supply interruption scenarios are not necessarily limited to the Middle East. Much of our imported oil now comes from Mexico. But even that friendly country is not a totally safe source; one could envision a sequence of events in which Central American guerrillas seize control of Mexican oil fields. A more real threat from Mexico, however—and other friendly suppliers such as Indonesia—is that they will reduce their oil exports in the face of increased domestic demand and dwindling reserves. Mexico, as a matter of fact, has already announced that it will have to cut down on exports *and* increase proven reserves just to keep up with growing domestic requirements through the 1990s. The Mexican Petroleum Institute says that unless new crude deposits are found, Mexico will not be a major oil exporter by 2000.

In *The Greenhouse Effect* I used the following analogy to describe what I thought the United States was doing to itself relative to energy supplies: "To be perfectly blunt about it, we probably are about to hang ourselves. The United States is standing on the gallows, the noose around its neck, just waiting for someone to yank open the trap door. If the greenhouse threat alone is not enough to trigger an urgent national resolve to kick the fossil habit and develop clean, renewable energy sources, the additional threat of being strung up by an empty gasoline hose should be." Well, we got a stay of execution. And we frittered away over a decade of opportunity (chapter 6).

We're still on the gallows.

Human Nature and Wisdom Teeth

I have to admit I've become much less of an optimist since I wrote *The Greenhouse Effect*. Over ten years have come and gone and the United States is still as reliant as ever on fossil fuels. Further, as I dig deeper into the global aspects of the greenhouse threat, the challenges begin to seem almost overwhelming. I'll explain why—and hope you'll pardon an aura of pessimism—in the next chapter of this book.

It's human nature, of course, to want to put off, or even avoid, dealing with unpleasant situations. Many years ago I had a couple of impacted wisdom teeth that would periodically give me problems. I'd chew on some aspirin and the ache would disappear, sometimes for months at a time. But the pain would always return. I knew what ultimately had to be done. But I also held out hope that just maybe the problem would be self-correcting. At any rate, human nature dictated that I wait until the pain became unbearable before I went to a dentist.

When the pain finally reached that level I was 8,000 miles from home in South Vietnam. The only dentist available was a young, relatively inexperienced air force officer who ended up literally *digging* out one of the teeth. After the "operation" he sent me back to my un-air-conditioned "hooch," where I lay on a cot for two days until the bleeding stopped. I wish I'd taken care of those teeth a lot sooner. It would have hurt a lot less.

And so it goes. We seem determined, as a nation, to wait until the pain becomes unbearable before reacting to the greenhouse threat and the impending energy crisis. We continue to harbor the unrealistic hope that the greenhouse effect isn't real, won't be so bad, might be reversed . . . and that maybe OPEC really has *our* best interests in mind. To return to the gallows analogy, we're going to wait until the trap door stars to drop. It's pretty tough to reverse things then.

But wait we will . . . for that $8.99 loaf of bread, for $3-a-gallon gas, or maybe just in service station lines.

And if you think technology can get us through drought and heat without significant crop and livestock losses, or that OPEC won't jack up prices and turn down the spigot, you've been doing a Rip Van Winkle for the last twenty years.

Going for a Rocket Ride

What happens when food prices soar, when the cost of gasoline surges, when we lose thirty-nine billion dollars annually in agricultural exports, when we shell out another forty-two billion dollars or more every year to OPEC? Inflation; and, as the trade deficit burgeons, probably "stagflation." If we're really unlucky, the crises of climate and energy will wound us simultaneously. If that happens we can forget the garden variety of inflation and experience a real treat: rocket-ride inflation.

Inflation is unwelcome for three primary reasons: (1) it erodes our money assets, (2) it leads to business bankruptcies, and (3) it has a tendency to accelerate.

While inflation may trigger booms in real-estate values or commodity prices, it eats away at our money assets. When prices rise, assets with fixed money value—such as savings accounts, cash value life insurance policies, and government bonds—lose worth. When inflation runs wild we can do nothing but watch helplessly as great chunks of our monetary cushions are swallowed by an endless cycle of price increases.

Inflation can also lead to widespread bankruptcies and business failures. The steady erosion of money assets makes it necessary for banks to charge ever-higher interest rates on loans to justify making the loans in the first place. Of course, borrowers are tempted to pay the high rates because they envision paying off the loans with ever-cheaper dollars. But this can make doing business terribly expensive for industries such as construction or utilities that rely heavily on borrowed funds to finance operations. New or poorly managed businesses can be pushed quickly toward insolvency.

Perhaps the greatest danger from inflation is its tendency to accelerate. The mere expectation of higher prices stimulates individuals and firms to buy more goods than they need (hoard) in order to stock up before prices jump again. But the very act of hoarding creates greater demand, which in itself leads to higher prices. So the whole process becomes self-fulfilling and the upward spiral of inflation quickens. As economist Robert Heilbroner points out, "Not everyone can march in the front rank in a parade, but the pace of a parade can change from a walk to a run if everyone tries to move up to the front rank." And although we haven't seen it yet in this country, we know the pace can change from a run to a rocket-ride, too.

In 1989, the inflation rate in Nicaragua soared to an astounding 10,000 percent. In Argentina it reached 3,079 percent, sparking rioting and looting. In 1991 the Argentine government instituted harsh economic measures in an effort to slow the hyperinflation. In 1990, Peru's inflation rate reached an out-of-control 7,650 percent. (In 1987 and 1988 the country was plagued by labor unrest and strikes; the upward spiraling prices also provided fertile ground for terrorist activities by the Maoist Shining Path guerrillas.) Brazil's inflation rate rocketed to over 2,900 percent in 1990, leading to severe economic recession and one of the largest foreign debts in the world. (U.S. consumer prices in 1990 climbed less than 6 percent.)

Stagflation

When inflation is accompanied by economic stagnation (or even recession and high unemployment) a condition known as "stagflation" results. Such was the case in the United States from the late 1970s to the early 1980s as inflation reached almost 14 percent (1980) and unemployment rates approached 10 percent (1982).

Economists refer to a trade deficit as an income "damper." That is, when more money is flowing out of a country to buy imports than is coming in from the sale of exports, there is a damper on income; income that could be used for business expansion and increased employment. The United States already has a huge trade deficit. So consider the effect on that deficit of losing thirty billion dollars in food exports. Consider the effect of shelling out another forty billion dollars (or eighty billion dollars?) every year to pay the ransom on increasing amounts of increasingly expensive foreign oil. Talk about an income damper.

So inflation soars, the economy stagnates, employment drops, crops and cattle struggle to survive . . . and people far away from here wonder where the food is. The stuff of a disaster movie? You bet. And we all have bit parts in this one. As I said earlier, *we* won't starve and *we* won't drown, but you can bet somebody will.

Who Was Draco?

Draco was an Athenian lawgiver and codifier who lived in the seventh century B.C. His code, of which only a small portion survives, was known for its harshness. Ever since, the word *Draconian* has stood for legal severity. It is a word with which we will become more familiar early next century, as the United States is forced to take exceptionally rigorous measures to deal with the greenhouse effect and address the energy crisis.

The reason they will be Draconian measures is because we will wait so long to respond to the crises. Then only extreme actions will give us a ghost of a chance of clawing our way back to normality. Even so we won't be able to undo what the greenhouse has already done, and we won't be able to avoid an extended period of stagflation or even recession. If we're lucky, perhaps we can cling to the edge of the pit without tumbling into the depths of a depression.

The measures, among other things, will be ones designed to enforce additional energy conservation (although higher petroleum prices in and of themselves will go a long way toward doing that) and to speed the development of alternatives to fossil-fuel-generated energy. They will be measures that should have been pursued through the 1980s and 1990s.

Conservation

Transportation currently burns two-thirds of all the oil used in this country, and automobiles account for 20 percent of our energy bill. By law, automakers' new car fleets were supposed to average at least 27.5 miles per gallon (mpg) by 1985. But there were delays in implementing that standard, and in 1988 U.S. Transportation Secretary James Burnley rolled back the average to 26.5 mpg. In 1989 the Bush Administration reestablished the 27.5-mpg figure effective 1990, but no new, higher standards were set for the future.

When the tough times come, stringent laws will establish fleet averages of at least 50 mpg along with a very short phase-in period. A gas-guzzler tax, currently imposed on automobiles averaging less than 22.5 mpg, will obviously be levied at a threshold much higher than that and be much stiffer than the existing levy.

And as if an OPEC-induced price explosion at gasoline pumps won't be bad enough, our government will be forced to add to the misery. It will do this by sticking on an additional price hike in the form of a tax of one to two dollars per gallon. And that may well be the smallest of the tax increases we'll see. The rapid development of alternative energy sources will require hundreds of billions of dollars, and you know where that money is going to come from.

Alternative Sources

Coal- and oil-fired electric utilities are the largest single source of fossil-fuel emissions in the United States. They account for 35 percent of all the CO_2 we put into the atmosphere. Thus they will be an obvious target for replacement with alternative generating sources, primarily nuclear and solar.

Like it or not, nuclear power plants will play a significant, perhaps even a leading role in our energy future. There will continue to be opposition to nuclear energy, of course, but such opposition will no longer be tolerated. (Even some of our "inalienable rights" will fall victim to the greenhouse effect.) Nuclear energy alone will not be the answer, though, and the solar energies—including wind, hydro, ocean, and biomass—will have to be greatly expanded.

It won't end there. Harsh rules will be legislated regarding insulation standards for new homes and efficiency standards for electrical appliances, such as water heaters, ovens, and refrigerators.

All in all, a scary scenario, I admit. We'll lie on the cot bleeding for quite a while.

It doesn't have to be that way, however. We've still got some time. A little time. But our actions through the 1980s make me less than sanguine about how we'll use that time.

Lester Brown, president of Worldwatch Institute, concerned over the greenhouse effect, ozone depletion, deforestation, and other environmental problems, agrees that we don't have many ticks of the clock left. "Time is not on our side," he says. "We have years, not decades, to turn the situation around, and even then there is no guarantee that we will be able to reverse the trends that are undermining the human prospect. But if we do, it will be during the nineties. Beyond that, it will be too late."

13/ Is There Any Hope?

Is there any hope that we will respond in a meaningful sort of way to the threat of a full-blown greenhouse effect before it reaches a crisis level? The answer is yes and no (well, what did you expect from a weatherman?) "Yes" from a standpoint that there is always *hope*. Certainly the scientific and technological challenges related to reducing greenhouse gas emissions are not insurmountable. "No" from the standpoint that we've already squandered more than a decade of opportunity, that it is human nature to ignore problems as long as possible, that our political system is geared to short-range perspectives (e.g., getting reelected in two, four, or six years), that the task of overcoming sovereign vested interests (politics) can be overwhelming, and that—heaven knows—we're already faced with a myriad of problems.

Some of the problems we face are truly life-and-death matters: AIDS and feeding the hungry, for instance. Other challenges are almost as critical: housing the homeless and combating poverty, to name two. Other important issues demand our constant attention: education, health care, national defense, the war against drugs, child welfare. Additional crises arise from our own folly, and as a result we find ourselves forced to spend billions of dollars in attempts to clean up our environment (even animals don't foul their own nests) and do such things as rescue the savings and loan industry (from itself). The Pogo Principle in full bloom.

Facing the Hurdles

The danger inherent in the greenhouse effect is that, left unchecked, it *will* press itself upon us in the form of an international life-and-death crisis. Not our death, to be sure, but certainly death within the Third World, and probably on a massive scale. This nation's life-and-death struggle, metaphorically speaking, will occur if the crises of climate and energy strike together. Then the United States will be forced to battle for its place as a world power in the face of monumental economic turmoil (chapter

12). Our economic strength, of course, is the very foundation of our global might.

But such a challenge is too hypothetical and too far in the future for us to contemplate. We are a nation conditioned to short-range views and tactical reactions, not to long-term thinking and strategic planning— except, perhaps, as it relates to military power.

Not all of our enemies carry guns and bombs.

The real impediments to action, then, are not scientific and techno- logical. They are psychological and political. Think of the scientific and technological aspects of the challenge as akin to running the 400-meter hurdles. Then think of the psychological and political facets of the problem as a twenty-six-mile (42-km) marathon . . . with hurdles every 100 meters.

Domestic Political Hurdles

Vice President Albert Gore understands the short-range perspective of our elected officials. "Congress has a bias against solving long-term problems," the Tennessean says. "The next election, the next budget cycle, the next redistricting: these are the sorts of concerns that command the most attention. An issue like global warming requires Congress to look decades ahead."

Gore has been one of the few political bright lights when it comes to dealing with comprehensive environmental issues. He recognized the greenhouse threat and cosponsored a bill designed to address it. The bill was introduced by former Senator Timothy Wirth of Colorado, another bright light. Wirth also understood the "twofer" principle outlined in the previous chapter: "I have read a lot of studies, met with a great number of people in this field, and have become convinced myself that the probability of a greenhouse effect having a significant impact is about 99 percent, and that what we ought to be doing is moving aggressively on programs of energy conservation, alternative energy sources, reforestation, and so on. And that even if that 99 percent is wrong . . . *those policies of reforestation, alternative energy programs, [and] energy conservation . . . are good for us anyway in terms of economic and environmental policy"* (italics added).

Gore and the Clinton administration are staring at a twenty-six-mile marathon with hurdles. In the last congressional session of the Reagan Administration three bills to encourage energy efficiency were intro- duced. None survived the first round of committee hearings.

The Reagan Executive Branch's response to Senator Wirth's bill was exemplified by the comments of the associate undersecretary for energy, Donna Fitzpatrick: "Scientific uncertainties must be reduced before we commit the nation's economic future to drastic and potentially misplaced policy responses."

Senator John Chafee of Rhode Island takes a more enlightened view when he says, "There are a great many questions about the greenhouse effect that can't be answered today. But I don't think we ought to let scientific uncertainty paralyze us from doing anything."

Stephen Schneider of NCAR gripes that for twenty years he's been hearing the same platitudes about scientific uncertainty used over and over as an excuse for government inaction. "Indeed," he explains, "how long we should study 'before' we act is not a scientific judgment but a value judgment, weighing the costs of any present investment to slow down the future climate change versus the costs of that change descending on us unchecked."

The government was still singing the same old scientific uncertainty song under President Bush. At a May 1989 multination meeting in Geneva, Switzerland, U.S. delegates argued that more study was needed before beginning work on an international treaty to reduce the impact and mitigate the effects of global warming. In November 1989, at an international conference in the Netherlands, the United States failed to support a proposal to freeze the level of emissions of greenhouse gases by the end of the century and cut them 20 percent by 2005. (But so did the Soviet Union, China, and Japan. And these nations, along with the United States, account for almost 60 percent of the world's output of greenhouse gases.) The United States agreed only that stabilizing emissions "should be achieved as soon as possible."

Although George Bush was elected under the banner of the "environmental president" in 1988, when it came to the greenhouse effect the banner was tattered and limp as the 1990s dawned. In addition to failing to back strong actions to address global warming, President Bush did not deliver on his campaign promise (such promises apparently are something less than real) to hold an international meeting on global warming his first year in office. He finally got around to it his second year in office, then angered European delegates to the meeting when he outlined U.S. policy on global warming. In a speech to the April 1990 conference Bush maintained that "a better understanding" of the science of climate change and the economics of dealing with it was needed before action could be taken. This was a variation on the theme Schneider had been hearing for two decades. Klaus Topfer, environment minister of West Germany, was one of those miffed at that attitude. "Gaps in knowledge must not be used as an excuse for worldwide inaction," he said. "The gravity of the situation requires immediate, determined action."

In 1991 the Bush Administration, at an international conference on climate change, submitted an "action agenda" to stabilize overall production of gases contributing to global warming. The plan did not, however, set targets for reducing CO_2 emissions, which would have been allowed to increase by 15 percent by the turn of the century.

In 1992, before the United Nations' earth summit (Conference on the Environment and Development) in Rio de Janeiro, Brazil, Bush said any commitments to reducing carbon dioxide releases must "fit each nation's particular circumstances." He went on to announce he could not accept "specific limits on CO_2 emissions." This left the United States alone among industrialized nations in opposing an earth summit treaty with fixed targets and timetables for scaling back such emissions. Many countries, including those of the European Community and Japan, pushed for an agreement calling for the reduction of greenhouse-gas releases to 1990 levels by the year 2000. The United States, arguing that such specificity would harm industry and promote economic hardship, found allies in China and a number of Third World nations eager to exploit coal reserves and develop nascent manufacturing bases.

In the end, the U.S. position prevailed, not because it was widely acceptable but because Bush said he would not sign an agreement with firm goals. The resultant eviscerated treaty called for industrialized countries merely to report periodically on their progress in curbing carbon dioxide and other greenhouse-gas releases, "with the aim of returning [them] . . . to their 1990 levels. . . ." Thus the Bush Administration's record of inaction on the greenhouse effect remained intact: without specific goals or targets there is no commitment; without commitment there is no action.

Psychological Hurdles

Schneider, when he talks about scientific judgments vis-à-vis value judgments, is absolutely right. The trouble is that it seems as though our value-judgment horizon rarely extends beyond our television sets (or, in the case of politicians, beyond the next election).

Former Senator Robert Stafford of Vermont, who introduced his own greenhouse legislation before retiring, had it figured out. Referring to the greenhouse effect, he said, "When a majority of the American public decides that something needs to be done . . . *it may take 20 years of discomfort* [italics added], but . . . sure, I think the human race is capable of beating this thing."

Capable, yes. Willing . . . I don't know. Humans are allegedly the most intelligent creatures on Earth. And yet, in the face of irrefutable evidence that smoking can be deadly, millions of Americans continue to puff away. In 1985, almost 400,000 of them died from smoking-related diseases. I find this a bizzare manifestation of intelligence.

Wearing seat belts in automobiles can save lives. In 1990, seat-belt laws were credited with holding the traffic fatality rate to its lowest level in at least two decades. And yet we continue to hear sophomoric rationalizations about not wearing seat belts, and indeed, as of 1991 only 59 percent of all drivers were strapping in. (But at least the percentage has been increasing.)

We don't always do what's good for us, even in the face of logic. You have to wonder about our resolve to address our energy future before the heat starts and the oil stops.

Despite my personal pessimism, there are signs of hope, not at the federal or personal level but at the state level, of all places! For example, New York has set a goal of reducing CO_2 emissions 20 percent by 2010. This is to be accomplished by mandating more energy-efficient buildings, automobiles, and appliances and by encouraging the use of alternatives to fossil fuels. New Jersey's global climate initiative targets the reduction of greenhouse-gas emissions through energy conservation and the plant-ing of trees.* Vermont, along with Irvine, California, and Newark, New Jersey, has voted to ban the use of CFCs. Two large California electric utilities, Southern California Edison and the Los Angeles Department of Water and Power, have announced they will reduce their CO_2 releases by 20 percent over the next twenty years, mainly by encouraging energy conservation.

Action is also being taken at the city level. Portland, Oregon, has drawn a boundary around the city beyond which growth is banned. Additionally, Portland is increasing the price of parking spaces and limiting their number in an effort to encourage car pooling and the use of mass transit. Denver, Colorado, is requiring its municipal fleet of

---

*It has been suggested that planting trees might be an answer to the greenhouse effect, because trees absorb carbon dioxide in the photosynthesis process. All we need is an area the size of Australia and we're home free. At least that's what the calculations of Gregg Marland of Oak Ridge National Laboratory in Tennessee suggest. In order to soak up the six billion metric tons of carbon put into the atmosphere each year by our burning of fossil fuels, we'd have to cover an entire continent with new plantings.

automobiles to exceed the federal corporate average fuel efficiency (CAFE) standard.

So, while the Bush White House continued to "study" the greenhouse effect, state and city grass-roots movements to actually do something about it started.

International Hurdles

But if the hurdles on a psychological and national level seem daunting, they appear virtually impassable on an international scale. For instance, consider the call by attendees of the World Conference on the Changing Atmosphere (Toronto, Canada, 1988) for creation of a world atmosphere fund to finance the development of technologies designed to reduce greenhouse-gas emissions. There was general conference agreement that such a fund was needed, but sharp differences of opinion regarding how the fund should be financed.

The final conference statement asked that the source of the funding be largely from a levy on the fossil-fuel consumption of industrialized nations; it also emphasized that much of the money would be used for applications within lesser-developed countries. A sizable minority of the conference participants (forty-eight nations were represented) questioned the wisdom of relying on taxes from industrialized countries for two reasons: (1) it would trigger significant national political resistance, and (2) fossil-fuel emissions aren't the only heavies in the greenhouse-effect drama and therefore should not bear the burden of financing the fund.

Certainly the nations of the former Soviet Union and Warsaw Pact, facing enormous economic struggles, can't be counted on to pitch in to the fund. In fact, they are confronted by overwhelming tasks in merely reducing their own CO_2 emissions—which account for a quarter of global emissions—and cleansing their hyperpolluted environments.

At least within the scientific and research communities international cooperation is forging ahead. In 1988 the Intergovernmental Panel on Climate Change (IPCC) was established by the World Meteorological Organization and the United Nations Environmental Programme. The IPCC mandate is to (1) assess available scientific information on climatic warming, (2) examine environmental and socioeconomic impacts of climatic warming, and (3) formulate response strategies.

On a political level, any attempt to actually curb CO_2 emissions is likely to be met with significant resistance from Third World countries. They blame industrialized nations for the greenhouse warming and aren't about to see their own economies suffer for lack of cheap energy if there is an attempt to constrain resources such as coal and oil. And this is where hoping for international concurrence on climate warming solutions gets scary. It's related to population growth.

Corpus Christi and Mexico

Every day, *every single day*, a net of over a quarter of a million people are added to the earth! That's equivalent to the population of Corpus Christi, Texas. Imagine, ninety to ninety-five million new individuals every year demanding food, shelter, and . . . energy. Over ninety million; that's the population of Mexico. True, most of the population explosion is occurring in the Third World, where energy demands aren't so great as in industrialized nations; but the demand is there all the same. And it's going to be satisfied by the cheapest, most available sources: wood, coal, and oil.

Although Third World demand for fossil fuels currently may be small relative to that of industrial powers, it's growing. For example, the global portion of fossil fuels used by the United States and Canada decreased from about 45 percent to 25 percent between 1950 and 1985. Meanwhile, Asian usage burgeoned from just over 1 percent to 11 percent, and the contribution from all developing nations jumped from 6 percent to 15 percent.

Most of the increase in Asian consumption can be accounted for by China, with one-fifth of the world's population. China is crammed with over a billion souls, and despite what Western democracies would consider repressive family planning policies, by the turn of the century China will likely be overflowing with 1.3 billion people, 100 million more than the national target.

China's economic goal is to triple its gross national product. To do this, vast amounts of energy will be required, much of which will come from China's extensive coal reserves. Over the next twenty years the nation is planning to essentially double its coal bill. And coal, of all the fossils, is the most prolific carbon dioxide producer.

The Solution

The solution to the greenhouse dilemma, and a few others, isn't difficult to figure out; but for reasons previously cited, it may be virtually impossible to execute. The solution is to reduce greenhouse-gas emissions.

The 1988 Toronto climate conference called for a 20 percent reduction in global carbon dioxide emissions by 2005. Senator Wirth's proposed legislation targeted a 20 percent reduction in U.S. CO_2 emissions over twenty years. (The bill also set goals for improved energy efficiency, sought increased funds for research into nonfossil energy sources, and encouraged U.S. efforts to coordinate initiatives to control global population growth, save tropical forests, and promote energy conservation.)

While reducing CO_2 emissions is an obvious and laudable goal, 20 percent isn't enough. A 50 to 60 percent cut is needed merely to stabilize atmospheric carbon dioxide concentrations. To be fair, however, a 20 percent rollback is probably the most we can reasonably hope for, and even that goal may be a pot of gold at the end of a rainbow.

Leaders like Wirth (now retired from Congress) and Gore understand this, but at least they recognize the seriousness of the greenhouse threat, and at least they've addressed it. Senator Chaffee appears to understand the cost of inaction, as do Senator George Mitchell of Maine and Representatives George Brown, Jr., of California and Bill Green of New York. Paul Tsongas, a 1992 presidential hopeful, realized over a decade ago the need for the United States to alter its energy course. Of my 1980 book he wrote: "*The Greenhouse Effect* analyzes one of the major reasons why we must speed the transition from fossil fuels to energy efficiency and renewable resources. It challenges us to go from understanding to action before time expires."

More than ten years later the challenge remains.

Conservation

The United States, despite energy conservation gains between 1973 and 1985, is relatively inefficient when it comes to industrial energy use. For example, we use twice as much energy as Japan, Germany, Denmark, and Switzerland to produce a dollar's worth of gross national product. There's a lot of room for improvement.

Arthur Rosenfeld, director of building sciences at the Lawrence Livermore Laboratory, University of California, sees the opportunity. He says that by investing 300 million to 500 million dollars in conservation measures over the next two decades, the United States could save up to

2.2 trillion dollars in energy costs, reduce its trade deficit, and cut back significantly its reliance on imported oil.

To achieve all of this, Rosenfeld suggests the following;

- Imposing annual increases in the gasoline tax, phasing in a dollar-per-gallon hike over ten years. This, he says, would encourage demand for fuel-efficient automobiles.
- Creating a scheme of penalties and rebates to be applied to the construction of new homes and commercial buildings; such a scheme would reward energy-efficient construction with lower electric rates and impose penalties on construction failing to meet specific energy standards.
- Developing a similar penalty–reward system for power plants to encourage electric utilities to reduce carbon dioxide emissions, improve energy efficiency, and invest in renewable energy sources, such as solar.
- Authorizing higher return-on-investment rates for utilities having the best energy efficiency records.

Rosenfeld is on the right track, although I don't agree with him on the imposition of a gasoline tax. A related proposal frequently talked about for reducing our reliance on foreign oil is that of levying a tariff, nominally pegged at ten to eleven dollars per barrel, on imported petroleum. I don't agree with that, either.

Taxes and Tariffs

The theory behind the tariff proposal is that by effectively increasing the price of oil, incentives would be created for the U.S. energy industry to develop sources with which to replace imported oil. These replacement sources would include not only sources other than oil, but oil itself: it is argued that an import fee would revive the domestic petroleum industry by making it economically feasible to restart drilling rigs in places such as Texas, Oklahoma, and Louisiana. Additionally, an oil import tariff would encourage consumer conservation, as well as lead to a reduction in our trade deficit.

Opponents of the import fee counter that it would saddle U.S. industry and agriculture with excessive energy costs, making those businesses uncompetitive in world markets. This could, in turn, cost more jobs than an import tax might save. A report issued by the Citizen-Labor

Energy Coalition, an advocacy group, states that a ten-dollar-per-barrel tariff could cost the U.S. economy 100 billion dollars a year. Opponents of the foreign oil fee also point out that the tax would not significantly hurt Arab OPEC nations but would more heavily penalize close U.S. allies who supply the majority of our imported oil: Mexico, Canada, Venezuela, Great Britain, Indonesia, and Nigeria.

Finally, we are warned that an import fee would lead to a policy of "drain America first," so that when world supplies do get tight—as they inevitably will—U.S. sources will have dwindled and we will be more heavily dependent than ever on foreign petroleum.

Assuming we take other action at this time, an oil import tax does not appear to be in our current best interest. Artificially induced higher energy costs would force U.S. industry into an even weaker position vis-à-vis other free world economies, especially those of already-strong Japan and Germany. The effect on our trade deficit could be devastating.

Further, it makes sense to hold on to U.S. oil reserves until they are really needed. When global supplies get tight, prices will rise naturally, and in turn make additional domestic drilling viable again.

To return to Rosenfeld's recommendation, a tax levied directly at gasoline pumps would be too regressive. That is, a dollar-per-gallon levy would be mere pocket change for persons who are well off, while those operating on low incomes or near the poverty level and needing an automobile to get to work would suffer significantly from such a tariff.

A Conservation Bonanza

There is a better alternative to imported oil fees and gasoline taxes. And that is to drive existing fuel-efficient technologies for automobiles into the marketplace through legislation. Autos account for 20 percent of our energy tab and thus can provide a conservation bonanza for the United States.

Deborah Bleviss, executive director of the International Institute for Energy Conservation, reports that automakers, led by European companies, have already developed a number of highly fuel-efficient prototypes. She says that "their technical advances could be used in automobiles and small trucks without sacrificing safety, comfort, performance, low emission levels or affordability."

Bleviss points out that Volvo, Renault, Volkswagen, General Motors, and Ford have working prototypes that get 60 mpg or more in city driving and 70 mpg or more on the highway. Renault's VESTA2, she says, gets 78 mpg in urban traffic and 107 mpg on the open road. "The car is tiny, seating only four people, but most of the innovations it incorporates could be scaled up to larger cars."

And already there are prototype electric cars that can reach top speeds of 70 mph (112 kph) and travel up to 150 miles (240 km) before requiring a battery recharge. Nissan has developed an electric car that can be recharged in just fifteen minutes. We will see more and more electric autos in the near future, for California has already ruled that by 2003, 10 percent of all new vehicles sold in the state must be "zero polluters." This virtually mandates electric cars, and other states are likely to follow suit.

Unfortunately, electric autos aren't a panacea for the greenhouse effect, since they will have to rely on electric utilities, most of which burn fossil fuels, for battery recharges.

What's Good for the Country Is Good for General Motors

U.S. car makers are understandably reluctant to bring smaller, more fuel-efficient cars into showrooms when they can make big profits selling big cars during a time of petroleum abundance. They need some "encouragement." And that encouragement must come in the form of a ten-year phase-in period during which the 27.5-mpg CAFE standard for new car fleets is raised to at least 40 mpg, as advocated by Senator Richard Bryan of Nevada. Additionally, the gas-guzzler tax threshold must be elevated to at least 30 mpg over the phase-in period. Further, the tax should be stiff, ranging from $1,000 for autos rated just under the threshold to $10,000 for cars found to be substantially below the standard.

Bleviss urges that we also consider instituting a program of government rebates for purchasing "gas sippers," cars that get significantly better mileage than the standard. As far as research and development goes, she says, we need a revamped federal program that incorporates "the automakers, their suppliers and small, innovative engineering companies, probably in an arrangement in which they share the costs of research." That is how it is done in Europe.

An argument often heard against encouraging "gas sippers" is that since such cars will be smaller, the downsizing will result in thousands of additional lives lost in traffic accidents every year. It's a specious argument. Fuel-efficient cars don't necessarily have to be smaller. Air bags can be made standard equipment in all vehicles. One hundred percent, instead of 59 percent, of drivers can use seat belts. We can perhaps learn that it is not speed per se that kills, but driving too fast for conditions. And most important, we can pay more than just lip service to getting drunk drivers off the road.

The U.S. auto industry certainly won't be an advocate of higher mpg standards. They will maintain they can respond to economic and market pressures as they evolve. What they don't realize is that the pressures will not "evolve" (which implies some sort of gradual change); they will explode. Crises, by definition, do not happen gradually. The climate and energy crises will not *ease* into place. That's the core basis of the argument for taking actions now.

U.S. automakers can wait until the crises strike . . . and then be left in the dust again as they were after the oil price shocks and embargoes of the 1970s. Foreign manufacturers moved in and cornered the market with their fuel-efficient cars then, and they'll do it again. The message here, I believe, is—to turn around an old quote—"what's good for the country is good for General Motors."

And while improving the mileage efficiency of cars can play a major role in getting the imported-oil monkey off our back, on a global scale and in terms of the greenhouse effect, we're perhaps fighting a losing battle . . . unless we switch to vehicles that do not pollute or that use nonfossil fuels. The problem is that there are 540 million vehicles in the world now, and by early next century that number will grow to almost a billion. Concomitant with that growth will come a 50 percent jump in CO_2 emissions. A report by James MacKenzie of the World Resources Institute and Michael Walsh, an independent transportation consultant, offers this assessment: "The daunting environmental problems facing manufacturers and policymakers will require a revolution in technology and thinking as profound as the one that heralded the advent of mass-produced mechanized transportation early in this century."

Automobiles aren't the only targets for energy conservation, of course; there are many opportunities for energy savings and efficiency at the consumer level. The federal government can stimulate consumer conservation by reinstating programs to offer low-cost loans, tax rebates, and tax credits for such efforts as adding home insulation, installing storm doors, and using solar energy. Legislation must be passed to mandate stringent energy-efficiency standards for new electrical and gas appliances. Guidelines could be established for the labeling of products whose manufacture and use does not harm the environment.

Changing Sources

Obviously, conservation alone isn't the answer. Conservation can reduce CO_2 emissions but not eliminate them. Elimination will require sources of energy other than fossil fuels. It is unrealistic, however, to think that an energy metamorphosis can take place quickly. History has shown that it requires about fifty years to switch from one type of major energy source to another. In view of this lengthy lead time, government-backed research and development programs aimed at alternative energy sources, particularly renewable ones (i.e., solar), must be immediately expanded by an order of magnitude. It is imperative that we ensure our energy future with sources that are inexhaustible . . . solar and its derivates.

However quickly we might take action, though, we will be forced to live with some undesirable alternatives for a while. Fossil fuels, undesirable from an environmental standpoint, won't disappear overnight. But because of their environmental undesirability, they must play a rapidly diminishing role in energy generation.

Natural gas is the cleanest of the fossils in terms of carbon dioxide, and it can be an important interim energy contributor. In the United States, natural gas currently supplies about one-quarter of our energy needs, but unfortunately we're using more than we're finding. Members of the oil and gas industry argue that further price decontrols are needed to stimulate retrieval of natural gas from many easy-to-produce resources. Still, natural gas production will continue to dwindle, and this relatively clean source of energy does not seem destined to become a major force in our energy future.

Ironically, coal, with its great abundance in countries such as the United States and China, is the filthiest of the fossils when it comes to producing greenhouse-gas and acid-rain (sulfur dioxide) emissions. The combustion of coal releases 75 percent more CO_2 to the atmosphere per equivalent unit of energy than does natural gas and up to 30 percent more than does oil.

Technologies are available to reduce sulfur dioxide release during coal combustion, but the best hope for reducing CO_2 emissions lies in developing a new, highly efficient "fuel cell." Fritz Kalhammer of the Electric Power Research Institute warns, however, that the technology for such a cell "isn't very highly developed." It will be a decade before electric generation plants using fuel cells can be built.

Nuclear Power

That brings us to another undesirable option: nuclear power. Nuclear power is undesirable not from a technical standpoint but from an emotional standpoint. As Senator Frank Murkowski of Alaska says, "nuclear power generation . . . is something that our country has got a phobia over."

Since 1978, no new nuclear plants have been ordered in the United States. After the accident at Three Mile Island, fifty-three plants were canceled between 1980 and 1984. The nuclear industry suffered another telling setback in 1986 with the Chernobyl power plant disaster in the Soviet Union. Although the Chernobyl facility was of a more primitive design than U.S. generating plants, Behran Kursunoglu, a nuclear advocate and theoretical physicist at the University of Miami, gloomily concludes, "Chernobyl was the ultimate dethroning of nuclear energy." He predicts many years will pass before atomic energy begins to regain political acceptance.

But we don't have many years. Our demand for electricity continues to increase. Since 1973 it has escalated by over 45 percent, and it will jump at least another 30 percent by the end of the century. New generating facilities will be required, and nukes currently offer the only available non-CO_2-emitting technology that can produce electricity on the immense scale required. France, which generates almost 75 percent of its electricity from nukes, calculates that this prevents over 120 million metric tons of carbon from being blown into the atmosphere each year.

The generation of electricity using solar energy—photovoltaic production—is not now a commercially viable answer to large-scale demand. Neither is nuclear fusion. Nuclear fusion is a nonradioactive process used to derive nuclear power. The process involves the fusing of two hydrogen isotopes into helium, the same element that powers the sun. One hydrogen isotope, deuterium, is available in seawater, and the other, tritium, can be manufactured. But despite continuing research, a commercial fusion reactor may not be available for another thirty to forty years.

Commercial electric generation from nuclear power (nuclear fission) in the United States has not resulted in a single death or injury to the public. But that does not mean there are no problems to be solved. Perhaps the most vexing is the disposal of nuclear waste, but that challenge is being worked on. Compared with other kinds of waste, the total amount of nuclear waste produced is quite small. In the United States, the tonnage of toxic chemical waste churned out each year is over 17,000 times greater than the nuclear waste generated over the past thirty years. In the future, the U.S. Department of Energy plans to seal spent solid nuclear fuel in corrosion-resistant canisters. The canisters will then be packed in absorbent clay in specially selected deep underground caverns . . . geologic formations that have been stable for millions of years. Both the National Academy of Sciences and the Congressional Office of Technology Assessment have endorsed this basic plan. Additionally, thousands of independent studies conducted worldwide have concluded that the technology for safe disposal of nuclear waste is available.

Other nuclear power plant issues are being addressed, too. The training given to operational personnel can be (and has been) markedly improved; the quality of new design and construction can be more closely monitored; and new plants can be located away from populated areas (this addresses an emotional issue, not a technical one). Indeed, we may need a whole new generation of reactors to satisfy the critics of nuclear power. This is something that Senator Wirth recognized and for which he requested support in his "greenhouse-effect" bill.

As Wirth himself says, we can't have it both ways. Given that we have to act now, we can't have both a nuclear-free world and a greenhouse-effect-free world.

Consider it in this light: for each future 1,000-megawatt nuclear plant we don't build, the alternative is a fossil-fuel-fired utility burning either 51,000 tons of coal each week or eight million barrels of oil per year.

Twenty-Gauge Politics

Beyond domestic energy conservation initiatives and efforts designed to develop fossil-fuel alternatives, there are actions the U.S. can take within the international arena to slow our trek toward a warmer world. We must play a sincere and highly visible role in preserving, and even restoring, the world's tropical rain forests. This may involve cutting off international development loans to countries that do not cease decimating their woodlands, or it may require committing manpower and equipment to special security forces organized to stop the destruction of forests. This is all easier said than done, of course. Remember, to a great extent the destruction of the world's rain forests is being carried out by the poor and the hungry, who are merely seeking a way to survive. There are more sinister forces at work, too.

Francisco Mendes Filho (Chico Mendes), a Brazilian rubber tapper by trade, organized other tappers to fight the rape of Brazil's rain forests. Sometimes he, his family, and his fellow workers would literally sit in the path of heavy equipment that was illegally clearing trees. Mendes gained international stature as an environmentalist and became known as the Amazonian Gandhi. He also became known as a nemesis to area cattle ranchers and land speculators who were responsible for much of the cutting. Death threats and attempts on his life were made. He was finally assigned police protection, but in the end it didn't help.

In late 1988, as he stepped outside his home for a backyard shower in the evening heat, Mendes was assassinated with a pointblank blast from a twenty-gauge shotgun. He staggered back into his house and died on the floor of his bedroom.

The Mendes tragedy might be more understandable if significant amounts of beef were being produced by Amazonian cattlemen on the cleared land. But this is not the case. The ranchers there are not motivated by profits derived from feeding people. They are driven by profits realized from billions of dollars of government subsidies secured by showing "productive use" of the land. And the cheapest way to show "productive use" is to clear and burn a tract, then let a few head of cattle wander around on it. Much of the acreage is held for speculation and even more profit.*

When there is so much money involved, neither human lives nor our environment will stand in the way of greed. On crossed phone lines—a frequent problem in the Amazon—one evening shortly after Mendes had been murdered, a caller overheard two men talking: "We got Mendes," a voice said. "Three to go."

And more did go. Although in 1990 a cattle rancher and his son were sentenced to nineteen years in prison for the slaying of Mendes, assassi-

---

*In early 1989 the Brazilian government temporarily stopped payment of its subsidies. But environmentalists argued it was too little too late. Later in the year, under pressure from the international community and domestic environmentalists, the government stepped up various conservation programs. It began a joint effort with Britain in forestry management and pollution control, accepted financial aid from West Germany to establish Amazon conservation programs, welcomed a gift of patrol aircraft from Italy and Canada for use in policing illegal timber cutting and burning, and approved a project backed by the United Nations and U.S. industrial firms to set up a huge protected forest region to produce rubber and nuts. In 1991, Brazilian president Fernando Collor de Mello promised to abolish permanently payment of the tax subsidies.

nations continued. In 1991, Expedito Ribeiro de Souza, a unionist and defender of the Amazon forest, was shot to death near his home. His death, as an activist, was only the latest in string of almost 1,600 dating back twenty-five years. Reverend Ricardo Resende, a Roman Catholic priest who presided at Ribeiro de Souza's funeral, thinks the assassination was part of a conspiracy engineered by large landowners who want to eliminate peasant demands for arable land.

After the funeral, Father Resende, also a land reformer, received a note saying he would be the next to die.

Such is the curious and confused web of human greed, tropical rain forests, and global warming. Life and death on our planet are lashed together in a complex and often obscure weave of environment, sociology, industry, politics, science, technology, and money.

But maybe most of all, money.

14 / Signs to Watch For

The debate over greenhouse warming could be—not will be, but could be—largely settled by the end of the 1990s or shortly thereafter. One argument heard repeatedly from greenhouse-effect skeptics is that since man has been injecting increasing amounts of carbon dioxide into the air for over 100 years now, we should have seen even more atmospheric warming than has occurred. These greenhouse doubters say the warming that has been measured is within the bounds of natural climatic fluctuation. We may well have evidence one way or the other by the turn of the century.

Global Temperature Trends

Certainly the most telling evidence will be in global temperature trends. Remember, the eight warmest years in over a century simmered the earth between 1980 and 1991. If the greenhouse warming proponents are correct, we'll see a resumption of record-breaking global heat during the latter part of the 1990s, after experiencing temporary cooling as result of Mt. Pinatubo's eruption. (You'll recall that because of Mount Pinatubo's legacy—volcanic aerosols in the upper atmosphere—global cooling was predicted by NASA's climate model for 1992 and 1993 [see chapter 7].) By the middle of the decade, if no other large volcano blows its top, worldwide temperatures should be off and running again.

Watch the media headlines. If the hot years return, as Hansen and others think they will, then by the end of the decade the rise in global temperature should be compelling enough to stifle most skepticism. The world will have warmed to levels not experienced for about 6,000 years (since the Post-Glacial Optimum, or Altithermal Period), and temperatures will have significantly exceeded any natural variation limits of the recent past.

Keep in mind, however, as pointed out earlier in this book, that global warming will not occur in a linear fashion. That is, it will not happen smoothly; there will still be yearly ups and downs in temperature. Every year will not be warmer than its immediate predecessor. The important thing to notice will be the frequency with which record warm years come along. If the greenhouse theorists are right—and, yes, I'm in their corner—the frequency of record hot years will be *at least* as great during the late 1990s and early 2000s as it was during the 1980s.

Watch and listen for the headlines.

Weather in the United States

In the United States, the pattern and frequency of heat waves, droughts, and hurricane activity over the next decade or so may well offer palpable evidence to many of us that we're crossing the Rubicon of global warming.

Record Heat, Record Cold

Using the Dust Bowl era of the 1930s as our analog, or model (see chapter 2), we can speculate about what weather "surprises" the 1990s might hold in store for us. One of the precursors to the unprecedented heat and dryness of the 1930s was the early emergence of record-breaking high temperatures in parts of the mid-Atlantic, mid-South, and southern states. In 1930, Delaware, Kentucky, Tennessee, and Mississippi all experienced sizzling heat reaching or exceeding 110° F (43° C). Thus temperatures of that magnitude in some states east of the Mississippi during the 1990s might be one of the earliest warnings of things to come.

As the hot spells of the 1930s became widespread, a hallmark of the anomalous weather was the extreme heat often experienced in July. The core of the hot weather hovered over the Dakotas, Nebraska, and Iowa, but a lobe stretched all the way through Georgia to north Florida (see figure 2.9). James Hansen's NASA climate model produced summertime temperature deviations similar to this for the latter half of the 1990s. But the NASA output under scenario A (see chapter 7) indicated an even larger area could be affected. The model output showed a broad region of summer heat, 2° to 4° F (1° to 2° C) above average, extending all the way from the Pacific Northwest eastward to the northern Plains, then southeastward through the Midwest and Deep South. (This is probably something Atlanta, Georgia, as it prepares for the 1996 Summer Olympic Games, doesn't wish to think about.)

Remember, over half the states established all-time maximum temperatures during the 1930s; virtually all of the readings reached or exceeded 110° F (43° C) (figure 2.3). A return of such heat during the 1990s would be a clear sign of how real the greenhouse threat is.

On a local scale, particularly during the latter part of the 1990s and into the early 2000s, you might be alert to how often hot weather comes along. For instance, there may well be an increase in the number of days on which the temperature reaches or exceeds a certain "hot" threshold, say 90° F (33° C) or 95° F (35° C). In many locations east of the Rockies, the frequency of such steamy weather is likely to increase by about half-a-dozen days per year (relative to "normal") after middecade.

But summer heat isn't the only anomalous warmth likely to occur concomitantly with global warming. Both the 1930s analog and the NASA climate model suggest winters, too, could be significantly warmer over much of the eastern and southern United States (see chapter 11).

This is not to say that cold waves would be a thing of the past. Quite the contrary (see chapter 5). The 1930s witnessed at least two periods of extended cold that are legendary. The first came in 1934, the same year the Dust Bowl dryness reached its peak. The 1934 frigidity began in late January as an arctic airmass poured into the northeastern part of the nation, setting the stage for the longest period of sustained cold ever to grip the region from Michigan to the Atlantic seaboard. The month of February 1934 turned out to be the coldest known in Boston, New York, and Buffalo. In Philadelphia, only January 1977 has been colder.

The second great icy period came during the winter of 1935–36. This time it was the northern Plains that shivered. In Langdon, North Dakota, the mercury failed to creep above 0° F (-18° C) for forty-one consecutive days during January and February. It was the coldest extended spell ever experienced in the forty-eight contiguous states.

In terms of the magnitude of the departure of temperature from normal, February 1936 was the coldest month on record in U.S. history. In parts of Montana, readings averaged 26° F (14° C) below normal. For Bismarck, North Dakota; Pierre, South Dakota; and Omaha, Nebraska, it was the most frigid month known. The winter as whole was the coldest ever for Pierre and Omaha, as well as for Minneapolis, Des Moines, and Kansas City, Missouri.

The great irony, of course, is that the historic cold of the winter of 1935–36 was followed, in the same region, by the unprecedented heat of the summer of 1936. That was the steamy summer that saw the mercury broil to at least 120° F (49° C) in the Dakotas, Kansas, Oklahoma, Texas, and Arkansas.

The Great Drought

In early 1992 we seemed a long way from the "Great Drought of the 1990s" (chapter 2). A mere 7 percent of the country was under the influence of severe or extreme drought (compared with 37 percent during the 1988 drought and 66 percent during the peak of the Dust Bowl in 1934). Out of sight, out of mind, I suppose. Yet it is drought, super-drought if I may be hyperbolic, that I feel remains the preeminent climatic threat to the United States during the 1990s.

As pointed out in chapter 2, a prediction of the precise timing of such an event is impossible. Still, as the decade marches on, the probability of a major drought, one rivaling that of the Dust Bowl, will increase steadily. Its arrival—not that we would look forward to it—would go a long way toward dispelling any doubt about the reality of the greenhouse effect.

What are the precursors, the signs, we can watch for? What might announce the advent of the Great Drought of the 1990s? To get one clue we can look at how the infamous Dust Bowl developed.

The initial probing attacks of the 1930s drought came to the Great Plains during the latter part of 1930. By early 1931, much of the Plains region was suffering severe to extreme drought. This marked the first peak of Dust Bowl desiccation. The parched region shrunk slightly in size during 1932 and early 1933, but after that the area of extreme dryness expanded with frightening rapidity. In 1934 and 1936 the entire region from Texas to Canada was scourged. The breadth of the drought waned again during 1937 and 1938 but exploded to one final peak in late 1939.

Thus one of the early hints of a developing major drought during the 1990s may be the growth of severe to extreme dryness on the Great Plains . . . particularly during the spring and summer, for as Harold Orville of the South Dakota School of Mines and Technology notes, "When dry conditions develop in the central United States in the spring and summer, the likelihood of their persistence for three to six months in the future is high."

Beyond that, there is a specific pattern of drought that we might watch for as a short-range precursor to the peak of widespread, major drought. It is a pattern that occurred just before the pinnacle of the droughts of the 1930s, 1950s, and 1970s.

In the springs of 1934, 1956, and 1977—the peak years of recent major dry spells—a broadly similar pattern of drought prevailed over the western two-thirds of the nation. Generally speaking, severe to extreme dryness gripped large parts of the Far West, Great Plains, and northwestern Great Lakes. The details of the pattern varied each spring; but frequently the most extreme desiccation stalked parts of Washington, Oregon, Idaho, Utah, Wyoming, and Colorado in the West, while on the Great Plains, sections of Minnesota, Wisconsin, Iowa, and Missouri were common targets. Admittedly, any drought is well established by the time this pattern emerges. But the pattern, considered hand in hand with Orville's statement, would suggest that things aren't about to get better. Far from it. It would warn of a summer, and perhaps an autumn, of intense, earth-cracking drought.

Keep in mind, though, that given the return of a superdrought, it isn't all dust and heat. Even in the midst of the great Dust Bowl, devastating floods thundered through parts of Colorado, Kansas, and Nebraska (1935). Sections of Texas (1936) and California (1938) saw homes and lives swept away by runaway rivers. And legendary floods roiled through New England (1936) and down the Mississippi and Ohio River valleys (1937). (See chapter 5 for a more complete discussion of these floods.)

Were Hugo and Andrew Harbingers?

In addition to clues garnered from heat waves and droughts in the United States, other signals that the greenhouse effect is a fact may come from the pattern and intensity of hurricane activity during the 1990s.

Again using the 1930s as a model, we should be alert for years with an abnormally large number of hurricanes and tropical storms in the Atlantic, Caribbean, and Gulf of Mexico. Although the 1930s did not produce a greater average annual number of storms than now occurs, the decade did feature the two "stormiest" hurricane seasons on record: 1933 with twenty-one hurricanes and tropical storms and 1936 with sixteen. The contemporary mean is ten, and the most active year in the last three decades was 1990 with fourteen tropical cyclones. Thus, if we are witness to a hurricane season or two during the current decade with the number of storms approaching eighteen or twenty, we can perhaps tally another signal for the reality of global warming.

But even more important than the number of hurricanes and tropical storms, we should be concerned over the frequency with which category three ("extensive" damage) or greater hurricanes (see chapter 3) rip across the coasts of the United States in the 1990s. During the 1930s, eight such storms thundered across our shorelines. By way of comparison, from 1970 to 1979 only four category three or stronger storms hammered the United States, while the 1980–89 period brought five such blows.

During 1930–39, most of the severe hurricanes that lumbered into the United States did so along the Gulf Coast or Florida shores. But no beaches along the Gulf or the Atlantic Ocean should consider themselves immune.

When Hurricane Hugo slashed into South Carolina in 1989, it had been thirty years since a major hurricane had pounded the Palmetto State's coasts. And when Andrew swirled through south Florida and Louisiana in 1992, it had been over a quarter of a century since those regions had seen a storm of such ferocity. Both Hugo and Andrew were category four ("extreme" damage) monsters that may well have been warnings of things to come. If global warming is increasing at the rate that Hansen and others have predicted, then there will be more Hugos and Andrews in our immediate future . . . more than the statistics of recent decades would suggest.

Passing Judgment

So we can watch and measure and analyze weather events of the 1990s. Table 14.1 summarizes the signs to watch for. Perhaps by the end of the decade the debate over global warming will be resolved. Perhaps it will continue. One thing is certain: we're good at analyzing and arguing; we're not so good at taking action. Few of us, for instance, have the resolve of Chico Mendes, the Brazilian rubber tapper whose assassination was detailed in the previous chapter. He had foresight and courage; I'm beginning to wonder if we have even hindsight. We've stepped on our environmental poncho so often it's beginning to look like our shredded ozone layer. We've fouled rivers so badly they've caught fire; we've polluted urban skies so severely they've changed color; we've dumped so much waste in the ocean, the oceans have spit it back on our beaches; we've pumped such a concentration of chemicals into the air that ecosystems and atmospheric layers have been eaten away; and we've buried so many chemicals in the ground, we've made entire communities uninhabitable.

Table 14.1. Global Warming Unchecked:
Signs to Watch For during the 1990s

Global Temperature Trends
- More record warm years during the latter half of the decade (after the cooling effects of Mt. Pinatubo's eruption disappear)

Temperatures in the United States
- Record heat with temperatures soaring to 110° F (43° C) or more in some states east of the Mississippi, but only as a prelude to . . .
- Sizzling heat on the Great Plains, with readings boiling to 120° F (49° C) or higher in some spots
- Locally, an increase in the frequency of "hot" summer days (mainly east of the Rockies)

Drought in the United States
- As a harbinger of a major drought: severe to extreme dryness developing on the Great Plains, probably during the spring or summer
- As a prelude to the peak of a major drought: severe to extreme dryness during the spring over much of the West (particularly parts of Washington, Oregon, Idaho, Utah, Wyoming, and Colorado) and parts of the Great Plains and northwest Great Lakes (particularly sections of Minnesota, Wisconsin, Iowa, and Missouri)

Hurricanes
- A few years with the number of hurricanes and tropical storms in the Atlantic, Caribbean, and Gulf of Mexico reaching eighteen or twenty (or more)
- More frequent hits by category three ("extensive" damage) or greater storms along Gulf and Atlantic shorelines (1992's Andrew provided a classic example: it struck *both* the Atlantic [south Florida] and Gulf [Louisiana] coasts with category four force)

James Webb, a former secretary of the navy and an accomplished novelist, wrote a commentary for *Newsweek* in which he reminisced over his daughter's transformation from a child to an adult and reviewed the legacy that his (my) generation had left her: "lacking in unity, riven by disagreement on every major issue and most minor ones, we have, perhaps unwittingly, encouraged our children to believe that there are . . . no commitments worthy of sacrifice." He ended his essay by saying, "we, the members of a creative, sometimes absurd, always narcissistic postwar generation, will soon receive a judgment. Whatever it is, our children have earned the right to make it."

I fear it will be a harsh one.

Yet we still have an opportunity to make a commitment, to be proactive instead of reactive for a change. We have a chance to finally cease ignoring the fact that we can (and do) irreparably alter our environment through our excesses and lack of vision. We can at last display a modicum of courage and respond to the multiple warnings and the sound—but admittedly not perfect—scientific predictions that say we're on the brink of warming our climate at a speed unknown in natural history. If this is not enough, we can at the very least embrace the concept of dual benefits and take positive steps toward our energy future. But it is the greenhouse effect that is currently most demanding and most palpable.

The wolf is in the door. Will we pretend not to see him . . . not to feel his breath . . . not to hear his footsteps?

Listen to what our world is telling us.

Glossary

Aerosols. Tiny solid or liquid particles found in the atmosphere.

Chlorofluorocarbons (CFCs). Chemicals, in varied forms—used as coolants and solvents, and in plastic food containers—that, under the influence of the sun's ultraviolet light in the upper atmosphere, undergo transformation and deplete ozone.

Climate analogs. Past climatic patterns that are expected to be analogous to future patterns, thus serving as guides to specific features of future climate.

Climate models. Complex mathematical mock-ups of atmospheric behavior run on large, high-speed computers and used to examine how our climate might respond to various, wide-ranging influences, such as increased carbon dioxide, the eruption of large volcanoes, etc.

Climatic stress. Weather patterns that lead to some regions becoming persistently—and abnormally—hot and dry at the same time others are cold and/or wet.

Dendrohydrology. The study of historical climates as revealed by the analyses of tree growth-rings.

El Niño **("The Child").** Originally a term applied to a warm ocean current that develops along the coast of Ecuador and Peru, usually just after Christmas. It is now the name given to a periodic, large-scale warming of the tropical eastern Pacific Ocean.

Evapotranspiration. Soil water loss from evaporation and the transpiration (giving off) of water vapor by plants.

Geritol solution. The postulation that if tons of iron were dumped into the world's oceans it would stimulate the growth of tiny algae which, in turn, would consume excess carbon dioxide.

Global warming. The long-term increase in average temperature at and near the earth's surface. Such warming is geographically uneven—some areas, in fact, can cool—but the overall net effect is a rise in global mean temperature.

Greenhouse effect. The overall warming of the earth's lower atmosphere primarily due to carbon dioxide and water vapor which permit the sun's rays to heat the earth, but then trap the heat near the earth's surface. The term often is used interchangeably with global warming.

Heat exhaustion. A condition that results from overexposure to heat or from too much activity in strong sunshine; characterized by pale, cold, moist skin, rapid pulse and breathing, low temperature, dizziness, and a tendency to vomit. Heat exhaustion is far less deadly than heat stroke, but death may occur in extreme cases.

Heat stroke. A condition that results from overexposure to heat—and often high humidity—or from too much activity in strong sunshine; characterized by a cessation of perspiration; dry, hot skin; rapid, irregular pulse; and a body temperature near 110°F (43°C). Coma, convulsions, and death may follow.

maf. A million acre feet of water—enough to cover an acre of land with water a foot deep. About 326,000 gallons.

Photosynthesis. The process by which plants combine CO_2 and water in the presence of chlorophyll and sunlight to form carbohydrates.

Photovoltaic production. A process that converts sunlight directly into electric power through the use of silicon crystal cells.

Smog. Originally a term applied to a mixture of smoke and fog. Today the term is commonly applied to largely urban air pollution which may contain, among other things, carbon monoxide, sulfur dioxide, and a variety of particulates.

Solar irradiance. Energy put out by the sun.

Westerlies. An undulating but generally west-to-east moving band of upper atmospheric winds that circle the hemisphere at mid-latitudes.

Index

HAROLD W. BERNARD, JR., received his degree in atmospheric science from the University of Washington. A past president of Boston-area chapters of the National Weather Association and the American Meteorological Society, Bernard is a member of the Climate Institute, a senior weather officer in the U.S. Air Force Reserve, and vice president of DBS Associates, Inc., and Impact Weather, Inc. He is the author of four other books, *The Greenhouse Effect, Weather Watch, The Travelers Almanac—North America,* and *The Travelers Almanac—Europe.*